Luiz Fernandes S. Dionisio
Antonio C. S. Lima

Monitoring oil palm pests in the south of the state of Roraima

Luiz Fernandes S. Dionisio
Antonio C. S. Lima

Monitoring oil palm pests in the south of the state of Roraima

Spatial distribution of Rhynchophorus palmarum (L.) and Metamasius hemipterus (L.)

ScienciaScripts

Imprint
Any brand names and product names mentioned in this book are subject to trademark, brand or patent protection and are trademarks or registered trademarks of their respective holders. The use of brand names, product names, common names, trade names, product descriptions etc. even without a particular marking in this work is in no way to be construed to mean that such names may be regarded as unrestricted in respect of trademark and brand protection legislation and could thus be used by anyone.

Cover image: www.ingimage.com

This book is a translation from the original published under ISBN 978-3-330-76594-8.

Publisher:
Sciencia Scripts
is a trademark of
Dodo Books Indian Ocean Ltd. and OmniScriptum S.R.L publishing group

120 High Road, East Finchley, London, N2 9ED, United Kingdom
Str. Armeneasca 28/1, office 1, Chisinau MD-2012, Republic of Moldova, Europe
Managing Directors: Ieva Konstantinova, Victoria Ursu
info@omniscriptum.com

Printed at: see last page
ISBN: 978-620-8-53668-8

Firstly to God, my fortress and my refuge.

To my beloved parents Francisco Hipólito Dionisio and Hortalina Alves Silva and siblings, for their love, dedication and unconditional support. Dedication.

Thank you

I thank God for the essence of experiencing the joys and overcoming the obstacles of this life.

I would like to thank the Federal University of Roraima for organising the course.

To the Coordination for the Improvement of Higher Education Personnel - Capes for granting the scholarship.

To Embrapa Roraima and all the collaborators for the logistics and support in carrying out the field work.

To Prof Dr Antonio Cesar Silva Lima for his friendship and teachings during the course of this work.

Dr Elisangela Gomes Fidelis de Morais for her co-supervision and for encouraging me to seek new knowledge.

To Prof Dr Paulo Roberto Silva Farias for his valuable teachings during my time at UFRA.

To the student Artur Vinícius Ferreira dos Santos for his statistical contribution and teaching.

To MSc. Ruy Guilherme Correia for his friendship, support, patience and encouragement in carrying out this and other work.

The Brasil BioFuels Company for making the study area available.

To my family for their support, friendship and affection.

My sincere thanks to everyone.

-Man is at constant war with nature and by defeating it he will be defeated"

(Luiz Fernandes Silva Dionisio)

SUMMARY

Oil palm (Elaeis guineensis Jacq.) is a palm of African origin typical of tropical regions, and in Brazil the largest cultivated areas are in the Amazon region. The aim of this study was to assess the spatial distribution pattern of Rhynchophorus palmarum and Metamasius hemipterus in an oil palm plantation in the south of the state of Roraima. The experimental area is located in the municipality of São João da Baliza, RR, where 24 samples were taken from September 2013 to August 2014. The perimeter of the area was demarcated with GPs, and the sampling grids were 45x45m. Spatial distribution was determined by placing a bucket trap with a food attractant (sugar cane) at each sampling point. Spatial variability and spatial dependence were analysed by incorporating geostatistical procedures based on spatial modelling techniques using semivariograms. Kriging maps were generated from R. palmarum and M. hemipterus adult count data. The sampling grids proved appropriate for characterising the spatial distribution of R. palmarum and M. hemipterus in the field. It was observed that R. palmarum infestation occurs both in the centre and at the edges of the plantation. The infestation of M. hemipterus begins at the edges of the plantation, where it is found in greater density and frequency. The spatial distribution of R. palmarum and M. hemipterus in the study area was aggregated, with spatial dependence described by the spherical model, forming "plots" of 100 *and* 210 metres and 78 to 199 metres in radius, respectively. The moderate spatial dependence index prevailed, occurring in 23 evaluations, and only one had strong dependence for R. palmarum. For M. hemipterus, there was moderate spatial dependence in 20 evaluations and only two with strong dependence. For R. palmarum the maximum range found was 210 metres with an area of influence of 13.9 ha and the minimum was 100 with an area of influence of 3.1 ha. For M. hemipterus the maximum range found was 199 metres with an area of influence of 12.4 ha and the minimum was 78 with an area of influence of 1.9 ha.

Keywords: Sampling; Coconut palm weevil; Oil palm; Geostatistics;

Pests.

SUMMARY

INTRODUCTION

The growing concern for sustainability is fuelling the development of renewable energy alternatives. One of these alternatives is the use of biofuels, which over the years has been included in Brazil's energy geopolitics, which justifies its incorporation into the agricultural production process in several Brazilian states, including the state of Roraima, which is developing oil palm cultivation for biodiesel production.

As in other regions of the Amazon, oil palm cultivation has the potential to be exploited in Roraima, especially in the southern region of the state, which has suitable soil and climate conditions for its adaptation (MORAIS et al., 2012). In Roraima, considering only the preferred areas, there are 406,121 ha suitable for utilisation, and the largest concentration of these areas is located in the southern region of the state (BASTOS et al., 2008; RAMALHO FILHO et al., 2010).

There are currently two companies based in the southern region of the state of Roraima growing oil palm. The Brasil BioFuels LTDA company, with a planted area of approximately 1800 ha, is based in the municipality of São João da Baliza, with plantations in the municipalities of São Luiz, São João da Baliza and Caroebe. Based in the village of Equador in the municipality of Rorainópolis, PalmaPlam LTDA has approximately 3000 ha of oil palm planted (personal information).

The oil palm (Elaeis guineensis Jacq.) or dendezeiro as it is known in Brazil, is a palm tree of African origin typical of tropical regions, and in Brazil, the largest cultivated areas are in the Amazon region. Long-term exploitation of this crop in the region can have positive environmental impacts, such as carbon sequestration in agricultural systems, recovery of degraded areas, as well as generating development, employment and income (SANTOS, 2008; CHIA et al., 2009; CORDEIRO et al., 2009; TAN et al., 2009).

Among the various types of oil palm, the only one that has acquired economic importance is the Elaeis genus, with the E. guineensis species being the most widely used in commercial plantations for agricultural exploitation, whose main product is the oil extracted from the pulp of the fruit, known as palm oil. It is also worth noting that this palm has a high oil production per unit area (VIANNA, 2006; BRAZILIO et al., 2012), when compared to other oilseeds.

Oil production from oil palm is approximately 10 times that of soya bean oil (Glycine max L.), double that of coconut oil (Coco nucifera L.), and around four times that of peanut oil (Arachis hypogaea L.) (CAVALIERO, 2003).

According to Costa (2007) the oil palm has a lifespan of between 20 and 30 years, and the crop produces from three and a half years after planting. The first planned plantings of this species took place in the state of Pará, from the 1960s onwards, allowing the expansion of the crop through private and public investment as part of regional development programmes (FAVARO, 2011). This experience was later transferred to the states of Amazonas, Amapá and Roraima, as well as Bahia (SANTOS, 2008).

Oil palm is a crop that is very susceptible to pest and disease infestation. Among the pests that infest this oilseed are rodents and insects, and among the diseases, the red ring (AV), fusariosis or lethal scale (SL), fatal yellowing (AF), marchitez Sorprevisa (MS), crown disease and dry heart rot (BRAZILIO et al., 2012).

In palm cultivation, phytosanitary problems related to insect pests have a significant economic impact (DUARTE et al., 2008), implying significant losses in productivity when it comes to a seed production field with considerable economic or genetic value (CYSNE et al., 2013).

Among the insect pests found in the crop are Rhynchophorus palmarum L. (Coleoptera: Curculionidae) and Metamasius hemipterus L. (Coleoptera: Curculionidae). These beetles make galleries in the apical meristem of palm trees, causing a reduction in tillering and the opening of the holes can serve as a gateway for phytopathogens (DUARTE et al., 2003; MOURA et al., 2006). One example is the nematode Bursaphelenchus cocophilus (Cobb.), the causative agent of red ring, a disease that can decimate up to 15% of a plantation (CHINCHILLA, 1992) and whose main vectors are these insects.

Considering that oil palm plantations have been expanding in the Amazon and especially in Roraima (RAMALHO FILHO et al., 2010), studies of the spatial distribution of R. palmarum and M. hemipterus in oil palm plantation areas could certainly contribute to good agricultural practices for use in the integrated management of these pests.

2 OBJECTIVES

2.1 General objective:

To determine the spatial distribution pattern and aggregation radius of R. *palmarum and M. hemipterus* in an oil palm plantation in the south of the state of Roraima defined by geostatistics.

2.2 Specific objectives:

2.2.1 Determine the model that best explains the spatial distribution of the insects *R. palmarum* and *M. hemipterus.*

2.2.2 Determine the aggregation radius of *R. palmarum* and *M. hemipterus* in oil palm plantations in Roraima.

2.2.3 Determine the number of traps/ha needed for reliable sampling of *R. palmarum and M. hemipterus.*

2.2.4 Determine the initial infestation sites of *R. palmarum* and *M. hemipterus* in oil palm plantations in Roraima.

3 LITERATURE REVIEW

3.1 Oil palm: general aspects

The oil palm (Elaeis guineensis Jacq.) has a simple trunk marked by scars. Its leaves are pinnate, large and have long leaflets, inserted at different angles, giving them a curly appearance and the petioles are made of fibrous tissue at the base and have spines on the margins. The male and female inflorescences are separated on the same plant and located in the leaf axils, with the male inflorescences having branches similar to hairy fingers. The fruits are arranged in clusters and are dense, ovoid, black and red at the base (MUSEU NACIONAL, 2014).

3.2 The economic importance of oil palm

The oil palm is native to West Africa and was introduced to the American continent at the beginning of the slave trade in the 15th century (TRINDADE et al., 2005). The oil palm crop stands out among oilseeds for its high oil production capacity per unit area. According to Murphy (2007) and Tinôco (2008), it is the oilseed with the highest vegetable oil productivity in the world, reaching up to 6 tonnes/ha, with yields exceeding 25 t/ha/year of bunches (CHIA et al., 2009).

The largest producers of palm oil are Indonesia, Malaysia, Thailand and Colombia with 53.73%, 32.64%, 3.61% and 1.64% respectively (USDA, 2013). The cultivated area in Brazil is approximately 170,000 hectares, with the state of Pará being the country's largest oil palm producer with over 90% of the planted area (SAGRI, 2013).

In Brazil, the oil palm was first introduced in the state of Bahia at the end of the 16th century and then in the Amazon region where the largest cultivated areas are currently located (BRAZILIO et al., 2012). The expansion of this crop brings many socio-economic benefits, generating development, employment and income for the Amazon region throughout the year, as production and harvesting are continuous, strengthening man's fixation in the countryside and preventing rural exodus (SANTOS 2008; CHIA et al., 2009; CORDEIRO et al., 2009). These benefits are possible due to the climatic characteristics of the Amazon region, which has conditions close to those required for oil palm cultivation (SALES, 2011).

3.3 Main insect pests in oil palm plantations

By domesticating a plant and turning it into a monoculture, a new ecosystem is formed that is different from the original, but these conditions favour the appearance of numerous insect pests.

According to Tinôco (2008) and Brazilio et al. (2012) the following insect pests stand out for attacking the oil palm: leaf-crown borer, Eupalamides cyparissias Fabricius, 1776 (Lepidoptera: Castniidae), leaf caterpillar, Brassoles *sophorae* Linnaeus, 1758 (Lepidoptera: Nymphalidae) and

Opsiphanes invirae Hubner, 1818 (Lepidoptera: Nymphalidae), palm weevil, Rhynchophorus palmarum (L.) (Coleoptera: Curculionidae), spittlebug, Metamasius hemipterus Linnaeus, 1758 (Coleoptera: Curculionidae), root borer, Sagalassa valida Walker, 1856 (Lepidoptera: Glyphipterigidae), which cause significant damage to the point of jeopardising the economic viability of the crop.

Among these insect pests, R. palmarum and M. hemipterus have been highlighted in recent years due to the damage they cause to palm trees and the fact that they jeopardise oil palm productivity, especially in the Amazon region.

3.4 Rhynchophorus palmarum (L.) (Coleoptera: Curculionidae)

R. palmarum, also known as the coconut weevil, is a polyphagous pest native to South America (ESPARZA-DIAZ et al., 2013) that has spread to the Middle East in recent decades (FALEIRO, 2006), and is considered an important agricultural pest in palm crops such as coconut and oil palm, as well as sugar cane, among others.

This pest is widely distributed in Latin American countries: Belize, Costa Rica, El Salvador, Guatemala, Honduras, Nicaragua and Panama (Central America), in Argentina, Bolivia, Brazil, Colombia, Ecuador, French Guiana, Paraguay, Peru, Suriname, Martinique, Trinidad Tobago, Uruguay and Venezuela (South America) and in India (EMPPO, 2007).

In North America the only record of R. palmarum was in Mexico (ESPARZADIAZ et al., 2013), however reports from the United States Department of Agriculture (2011) confirmed the presence of the palm weevil in San Diego in the United States.

In Brazil, this insect has been reported in several states, mainly in places that grow palm trees such as coconut (Cocus nucifera) and oil palm (Elaeis guineensis Jacq.). Thus, the main regions where this pest can be found are the North and Northeast.

R. palmarum belongs to the class Insecta; subclass Pterygota, order Coleoptera and family Curculionidae. It is characterised by having a rostrum with a mouthpart at the end. According to Triplehorn and Jonnson (2011) all the individuals in this family are phytophagous, feeding on both living and dead plants, attacking almost every part of a plant. The genus Rhynchophorus currently contains nine species, six of which are known to attack palm trees.

The adult insect (Figure 1) is a black-coloured beetle with a small head that elongates forward in the shape of a rostrum. It varies in size and can measure between 40 and 60 cm in length and has a diurnal and crepuscular habit. R. palmarum is an insect that undergoes complete metamorphosis, i.e. the egg gives rise to the larva which pupates and then reaches the adult stage. The adult can live for up to 60 days and during this time the female can lay up to 700 eggs. The male, when on the host plant, releases the aggregation pheromone which attracts both males and females, usually to the food source. Once both sexes are in the same environment, copulation also takes place. The females lay eggs in newly opened wounds on the plant, restarting the entire life cycle (LACERDA, 2001, SÁNCHEZ-SOTO; NAKANO, 2002, NAVARRO et al., 2002; RAMOS, 2012).

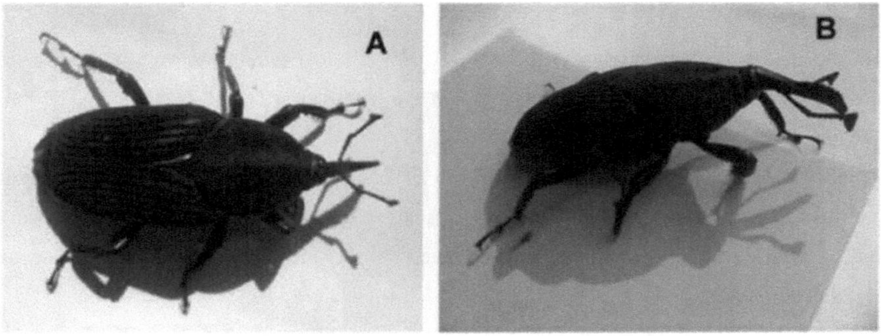

Figura 1. Rhynchophorus palmarum collected with traps in an oil palm plantation in the south of the state of Roraima. (A) female and (B) male. Photo: Luiz Fernandes Silva Dionisio

As well as causing direct damage, R. palmarum is the main vector of the nematode Bursaphelenchus cocophilus (Cobb) Baujard (GERBER; GIBLIN, 1990), which causes the red ring of the coconut palm. This disease is lethal to the coconut palm and other palm trees (SÁNCHEZ-SOTO; NAKANO, 2002; NAVARRO et al., 2002; COSTA-CARVALHO et al., 2011).

R. palmarum has been reported on 35 plant species from 12 different families, although it is predominantly found on Arecaceae. In this case, the main hosts, apart from the oil palm, are: Cocos nucifera (coconut palm), Euterpe edulis (juçara palm), Metroxylon sagu (sago palm), Phoenix canariensis (canary palm), Phoenix dactylifera (date palm), Saccharum officinarum (sugar cane). Other less significant hosts are: Ananas comosus (pineapple), Annona reticulata (true or countess fruit), Artocarpus altilis (breadfruit), Carica papaya (papaya), Citrus sp. (citrus fruit), Mangifera indica (mango), Musa sp. (banana), Persea americana (avocado), Psidium guajava (guava), Theobroma cacao (cocoa) (EMPPO, 2005).

3.5 Metamasius hemipterus (L.) (Coleoptera: Curculionidae)

Species of the genus Metamasius are found on all continents, mainly in the tropical region of the Americas. Approximately 110 species have been described worldwide, associated with various host plants such as arecaceae, bromeliads, musaceae, sugar cane, among others (ZORZENON et al., 2000; ROCHA, 2012).

Several species of Metamasius are reported as pests of economic importance in the state of Florida, United States, causing losses in pineapple, banana and sugar cane crops, as well as attacking native bromeliad species, orchids and ornamental palms (0'BRIEN; THOMAS, 1990; ZORZENON et al., 2000; SOLIMAN et al., 2009).

In Brazil, Metamasius is associated with several hosts: açaizeiro (Euterpe oleracea Martius), coqueiro (Cocus nucifera L.), dendezeiro (Elaeis guineensis Jacq.), palmito-juçara (Euterpe edulis Martius), pupunheira (Bactris gasipaes), tamareira (Phoenix dactylifera), Australian royal palm

(Archontophoenix spp.) all belonging to the Arecaceae (Arecales) family, as well as sugar cane *(Saccharum officinarum* Linnaeus) and banana *(Musa* spp. - Musaceae) (ZORZENON et al., 2000; YASUDA, 2005; GOMES, 2008).

The species M. *hemipterus,* commonly known as the 'spittlebug', is considered an important pest of palm trees, along with *R. palmarum*, because its larvae bore into the stem of the plant causing losses due to reduced tillering and the opening of holes can serve as a gateway to phytopathogens (SOLIMAN et al., 2009).

M. hemipterus lay milky-white eggs, elliptical in shape and about 2 mm long. The larvae are apodous, milky-white in colour, with a yellow to brown cephalic capsule and can reach a length of more than 8 to 10 mm in the last instars. The pupae are protected inside a cocoon made by the larva itself using fibres from the host plant (LEÓN- BRITO et al., 2005; ZORZENON et al., 2000).

In the adult stage, M. hemipterus is 10 to 20 cm long and is usually coloured red and black or orange and black (Figure 2). It has diurnal habits and great flying skills, and its cycle varies from 2 to 3 months, with the adults appearing at the hottest times of the year. The females scrape the stem tissue with their rostrums and lay their eggs, leaving them protected. The damage is caused by the larvae, which feed on the plant's vegetative tissue and form galleries along the stem (GALLO, 2002).

Figura 2. Metamasius hemipterus collected with bucket traps in an oil palm plantation in the south of the state of Roraima. Photo: Luiz Fernandes Silva Dionisio

3.6 Spatial distribution of insects

Studies on the spatial distribution of pest populations in commercial crops are of great importance in integrated pest management programmes as they help with the timing of control, the development of sampling plans and the planning of experiments in the field (RUIZ et al., 2003), requiring differentiated methods according to the particularities of each species (COSTA, 2009).

Once the spatial distribution pattern of an insect is known, it is enough to improve the sampling processes (BARBOSA, 1992; YOUNG and YOUNG, 1998; FERNANDES et al., 2003; STECCA, 2011), implementing integrated insect management and concentrating the sampling effort on the places where it is found (HASSEL, 1987; BARRIGOSSI et al., 2001; COSTA, 2009).

Dispersal in the habitat can occur in different ways between species and between populations of the same species, so that variation in the form of spatial distribution can occur as a result of environmental or genetic factors. Arrangement patterns can be determined using dispersion indices and theoretical frequency distributions (BARBOSA, 1992; YOUNG and YOUNG, 1998; FERNANDES et al., 2003) or using geostatistics (SILVA et al., 2011).

In order to study the spatial distribution of an insect, it is first necessary to define the size of each sampling unit, the number of sampling units and how these units are to be allocated in the experimental area (COSTA, 2009).

The spatial arrangement of insect populations follows patterns that can be classified as: random, uniform or aggregated (RICKLEFS, 2003). According to Taylor (1984), spatial distribution is an ecological characteristic of the species, explained by population dynamics as a product of the growth of individuals, such as birth, death and migration.

According to Barbosa (1992), for studies on insect distribution, there is a need to know the frequency distributions of the number of individuals of each pest species, adopting sampling criteria to estimate population parameters. The author also points out that, in addition to the mathematical models commonly applied, the use of geostatistics for analysis makes it possible to describe the spatial dispersion of pests, estimate the errors of population variables, verify the effects of environmental factors on population parameters and changes in populations over time and space (SILVA et al., 2011). Therefore, the use of geostatistics to determine the spatial distribution of insects has become increasingly common (GREGO et al., 2006; DINARDO- MIRANDA et al., 2007; FARIAS et al., 2008; SILVA et al., 2011; DAL PRÁ et al., 2011).

3.7 Using geostatistics to study the spatial dependence of insects

The methods commonly used to describe the distribution of a population allow distribution patterns to be classified into three categories: aggregate, uniform or random. However, these (mathematical) methods do not take into account the spatial location of the sampling points, which is why the use of geostatistics has been the most appropriate tool for studying the spatial dependence of a pest, as it takes into account the position in space of the variable under study (LIEBHOLD et al., 1993; LEAL, 2009), making it possible to quantify this dependence between samples collected in the field and, from the preparation of semivariograms that are adjusted to a model that provides the radius of aggregation of the species in the area, to construct distribution maps (LEAL et al., 2010).

The incorporation of geostatistical procedures based on spatial modelling techniques using semivariograms and estimation via kriging are important tools that can be used to determine, among other parameters, spatial distribution and to develop safe field sampling methods.

Several studies have been carried out using geostatistics to verify the spatial behaviour of pests of economic importance. These include Grego et al. (2006) who determined the spatial distribution of the wheat caterpillar (Pseudaletia sequax Franclemont) in the triticale crop (Triticum secale Wittmack) in the state of São Paulo; Dinardo-Miranda et al. (2007) on the distribution of the root leafhopper (Mahanarva fimbriolata) in sugar cane in the state of São Paulo; Farias et al. (2008) who characterised the spatial distribution of the cartridge caterpillar (Spodoptera frugiperda) in maize plantations in the state of São Paulo; Leal (2009) verified the dispersal of the Asian citrus psyllid (Diaphorina citri Kuwayama) in two citrus orchards in the state of São Paulo, Silva et al. (2011) observed the spatial distribution of the citrus black fly (Aleurocanthus wogllumi Ashby) in a citrus orchard in the state of Pará; Dal Prá et al. (2011) characterised the spatial distribution of pasture corm larvae (Diloboderus abderus) in oat crops in the state of Rio Grande do Sul.

3.8 Geostatistics in data evaluation

By analysing geostatistics, semivariogram models are fitted to the data obtained and, based on the most appropriate model, maps of spatial dependence are constructed (LEAL, 2009).

3.8.1 Semivariogram

The semivariogram provides information on the type and form of spatial dependence and, according to VIEIRA et al. (1983), is the first stage of geostatistics, providing data for making maps. It is a graph of the semivariance as a function of distance that characterises the structure of the spatial dependence of the variable under study, i.e. the semivariogram is a function that relates the semivariance to the distance vector.

The semivariogram is defined by three parameters: the plateau (C0 + C1), the nugget effect (C0) and the range (a). The plateau is the value of the semivariogram corresponding to its range, i.e. from that point onwards there is no longer any spatial dependence between the samples. The nugget effect is the semivariance at very close points, when the distances between the sampling units are very small; and the range measures the limit distance of spatial dependence.

Several researchers (GREGO et al., 2006, DINARDO-MIRANDA et al., 2007, FARIAS et al., 2008, LEAL, 2009, SILVA et al., 2011, DAL PRÁ et al., 2011) have found the spherical mathematical model to be the most suitable for describing the spatial distribution behaviour of insects. In this model the plateau and range are clearly identified and generally the nugget effect is small in relation to this plateau. Figure 3 shows an idealised semivariogram.

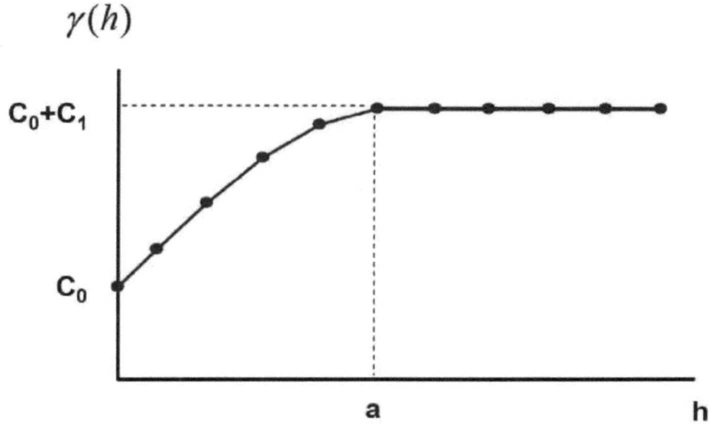

Figure 3 - Semivariogram fitted to the spherical model with the minimum semivariance parameters, nugget effect (C0), plateau (C0+C1) and range (a).

For spatially dependent properties, the difference between values *[Z(x) - Z(xi+h)]*, where the mean is, is expected to increase with distance up to a certain point, from which it stabilises at a certain value, a plateau (c1) and approximately equal to the variance of the data. This distance is called the range (a) and represents the radius of a circle within which the values are so similar to each other that they are correlated. The value of the semi-variance at the intersection of the y-axis is called the nugget effect (C0) and represents the variability of the property under study at smaller spacings than the one sampled. Thus, the greater the nugget effect, the weaker the spatial dependence of an attribute (VIEIRA et al., 1983). The semivariogram is estimated by:

$$\gamma\,(h) = \frac{1}{2\,N\,(h)} \sum_{i=l}^{N(h)} [Z(x_i) - Z(x_i + h)]^2$$

where,

γ(h) is the semivariance

N(h)is the number of observation pairs [Z(xi); Z(xi+h)] separated by distance h.

The model used will be the one whose function best represents the observed semivariance.

a. Spherical Model:

This model is obtained by selecting the values of the nugget effect (C0) and the plateau (C0 + C1), drawing a line that intersects the y-axis at C0 and is tangent to the first points near h = 0. This line will cross the plateau at a distance a' = (2/3)a. Thus, the range (a) will be a = 3a72, and this model is linear up to approximately (1/3)a (VIEIRA, 2000). Thus, the spherical model is defined as:

$$\gamma(h) = C_0 + C_1\left[\frac{2}{3}\left(\frac{h}{a}\right) - \frac{1}{2}\left(\frac{h}{a}\right)^3\right] \rightarrow 0 < h < a$$

$$\gamma(h) = C_0 + C_1 \rightarrow h \geq a$$

b. Exponential model:

The parameters C0 and C1 of this model are determined in the same way as for the spherical model (VIEIRA, 2000). However, the difference between the spherical model and the exponential model is that the latter reaches its plateau asymptotically, with a range of

practical defined as the distance at which the model value is 95 per cent of the baseline (ISSAKS; SRIVASTAVA, 1989).

$$\gamma(h) = 0 \rightarrow h = 0$$

$$\gamma(h) = C_0 + C_1\left\{1 - \exp\left[-\frac{3h}{a}\right]\right\} = h > a$$

c. Gaussian:

Similar to the exponential model, the Gaussian model reaches the plateau asymptotically, and the parameter is defined as the practical range or distance at which the model value is 95% of the plateau (ISSAKS; SRIVASTAVA, 1989). However, what characterises this model is its inflection point close to the origin.

$$\gamma(h) = 0 \rightarrow h = 0$$

$$\gamma(h) = C_0 + C_1\left\{1 - \exp\left[-3\frac{h^2}{a}\right]\right\} = h > a$$

d. Power:

The power model does not reach a plateau and is generally used to model phenomena with infinite dispersion capacity.

$$\gamma(h) = 0 \rightarrow h = 0$$

$$\gamma(h) = C + Ah^B \quad 0 < B < 2 \rightarrow h > 0$$

The parameter B must be strictly greater than zero and less than 2 in order to ensure that the semivariogram has conditional definite positivity.

After the adjustment, the kriging maps are constructed using the spatial dependence modelled in the semivariogram, estimating values at any position in the field without a trend and with minimal variance. Using these maps it is possible to see the behaviour of the variable under study.

3.8.2 Kriging

Kriging uses the spatial dependence between neighbouring samples, modelled in the semivariogram, to estimate values at any position within the field, with no trend and minimal variance (VIEIRA et al., 1983).

According to SILVA JUNIOR (2001), kriging can be called an interpolator because it weights the values of the points neighbouring the point to be estimated, obeying the criteria of non-bias and minimum variance. In this sense, kriging is nothing more than a weighted measure of the observed values of a given variable within a neighbourhood. Only points that are spatially related to the location to be estimated are used in kriging (VIEIRA et al., 1983).

In this case, in order to draw up the map to evaluate the study, it is necessary to interpolate the sampled points to obtain an estimate. This estimate (z*) is a linear combination of the values of the neighbouring measurements (x0), as shown in the equation below:

$$Z^*(x_0) = \sum_{i=1}^{N} \lambda_i Z(x_i)$$

in which:

z* = estimate,

x0 = linear combination of the values of the neighbouring measurements,

N = number of measured values involved in estimating z (xi), λ_i= weight associated with each measured value.

4 BIBLIOGRAPHICAL REFERENCES

BARBOSA, J. C. Sequential sampling. In: FERNANDES, O. A.; CORREIA, A. C. B.; DE BORTOLI, S. A. Integrated pest and nematode management. Jaboticabal: FUNEP, 1992. p. 205-211.

BARRIGOSSI, J. A. F.; YOUNG, L. J.; CRAWFORD, C.A.G.; HEIN, G. L.; HIGLEY, L. G. Spatial and probability distribution of Mexican bean beetle (Coleoptera: Coccinellidae) egg mass populations in dry bean. Environmental Entomology, v.30, p. 244-253, 2001.

BASTOS, T. X. MULLER, A. A. RAMALHO FILHO, A. PACHÊCO, N. A. GONSALVES, A. O. MONTEIRO, D. C. A. Agroclimatic and climate risk zoning for oil palm cultivation in the state of

Pará. Embrapa Amazônia Oriental: 2008. 27p. (Documents 319).

BRAZILIO, M.; BISTACHIO, N. J.; PERINA, V. de C. S.; NASCIMENTO, D. D. do. The oil palm *(Elaeis guineensis* Jacq.) - Review. Bioenergia em revista: diálogos, v.2, n.1, p.27-45, 2012.

CAVALIERO, C. K. N. Insertion of Regulatory Mechanisms to Incentivise the Use of Alternative Renewable Energy Sources in the Brazilian Electricity Sector and in the Specific Case of the Amazon Region. Thesis (Doctorate in Energy Systems Planning), State University of Campinas. 2003.

CHIA, G. S.; LOPES, R.; CUNHA, R. N. V.; ROCHA R. N. C.; LOPES M. T. G. Repeatability of bunch production of interspecific hybrids between kayak and oil palm. Acta Amazonica, v. 39, n.2, p. 249-254, 2009.

CHINCHILLA, C.M. El síndrome Del anillo-rojo hoja pequena en palma aceitera y cocotero. Revista Palmas, v.13, p.33- 55, 1992.

CORDEIRO, A. C. C.; MACIEL, F. C. da S.; SMIDERLE, O. J.; ALVES, A. B. Vegetative development of oil palm in savannah and forest ecosystems in Roraima. Boa Vista, Embrapa Roraima. Research and Development Bulletin, n.10, p. 16, 2009.

COSTA, M. G. Spatial distribution and sequential sampling of nymphs and adults of Diaphorina citri kuwayama (Hemiptera: Psyllidae) in citrus cultivation. 2009. 106f. Thesis (PhD) - Universidade Estadual Paulista, Faculdade de Ciências Agrárias e Veterinárias.

COSTA, R. C. Inventory of the life cycle of Biodiesel obtained from palm oil for the conditions of Brazil and Colombia. Dissertation (Master's in Energy Engineering). Federal University of Itajubá, 2007.

COSTA-CARVALHO, R. R. da.; WARWICK, D. R. N.; SOUZA, P. E. de.; CARVALHO FILHO, J. L. S. de. Longevity of Thielaviopsis paradoxa, causal agent of coconut resinosis in Rhynchophorus palmarum. Scientia Plena, v. 7, n. 4, p. 1-6, 2011.

CYSNE, A. Q.; CRUZ, B. A.; CUNHA, R. N. V. da.; ROCHA, R. N. C. da. Population fluctuation of Rhynchophorus palmarum (L.) (Coleoptera: Curculionidae) on oil palms in Amazonas. Acta Amazonica. v. 43, n. 2, p.197-202, 2013.

DAL PRÁ, E.; GUEDES, J. V. C.; CHERMAN, M. A.; JUNG, A. H.; SILVA, S. J. P. da.; RIBAS, G. G. Use of geostatistics to characterise the spatial distribution of Diloboderus abderus larvae. Ciência Rural, v.41, n.10, p. 1689-1694, 2011.

DINARDO-MIRANDA, L. L.; VASCONCELOS, A. C. M.; VIEIRA, S. R.; FRACASSO, J. V.;

GREGO, C. R. Use of geostatistics in the evaluation of the spatial distribution of Mahanarva fimbriolata in sugarcane. Bragantina, v.66, n.3, p.449-455, 2007.

DUARTE, A. G.; LIMA, I. S.; NAVARRO, D. M. A. F.; SANT'ANA, A. E. G. Capture of Rhynchophorus palmarum L. (Coleoptera: Curculionidae) in traps baited with the aggregation pheromone and volatile compounds from pineapple fruit. Revista Brasileira de Fruticultura, v.25, n.1, p.81-84, Apr. 2003.

DUARTE, A.G.; LIMA, I.S.; ARAÚJO JÚNIOR, J.V.; DUARTE, A.G.; ALBUQUERQUE, A.L.S.; CRUZ, M.M. Disposition of the nematode Bursaphelenchus coccophilus (Cobb) Baujard, in coconut trees with red ring disease. Revista Brasileira de Fruticultura, v.30, n.3, p. 622-627, 2008.

EMPPO - EUROPEAN AND MEDITERRANEAN PLANT PROTECTION ORGANISATION. Rhynchophorus ferrugineus and Rhynchophorus palmarum. France: EMPPO Bulletin, n.37, p. 571-579, 2007.

EMPPO - EUROPEAN AND MEDITERRANEAN PLANT PROTECTION ORGANISATION. Data sheets on quarantine pests, Rhynchophorus palmarum. France: Bulletin, n.35, p. 468-471, 2005. Available at:<http://www.eppo.org/QUARANTINE/insects/Rhynchophorus_palmarum/DS_Rhynchophorus_palmarum.pdf> Accessed on: 21 Apr. 2014.

ESPARZA-DÍAZ, G.; OLGUIN, A.; CARTA, L. K.; SKANTAR, A. M.; VILLANUEVA, R. T. Detection of Rhynchophorus palmarum (Coleoptera: Curculionidae) and Identification of Associated Nematodes in South Texas. Florida Entomologist, v. 96, n.4, p. 1513-1521, 2013.

FALEIRO, J. R. A review of the issues and management of the red palm weevil Rhynchophorus ferrugineus (Coleoptera: Rhynchophoridae) in coconut and date palm during the last one hundred years. International Journal of Tropical Insect Science, v.26, n.3, p.135-154, 2006.

FARIAS, P. R.S.; BARBOSA, J. C.; BUSOLI, A. C.; OVERAL, W. L.; MIRANDA, V. S.; RIBEIRO, S. M. Spatial analysis of the distribution of Spodoptera frugiperda (J.E. Smith) (Lepidoptera: Noctuidae) and losses in maize crop productivity using geostatistics. Neotropical Entomology, v.37, n.3, p.321-327, 2008.

FAVARO, F. T. Sustainable development in the Amazon: a new approach to oil palm with the utilisation of residual biomass for energy generation. 2011. 159f. Dissertation (master's degree in energy). University of São Paulo.

FERNANDES, M. G.; BUSOLI, A. C e. BARBOSA, J. C. Spatial distribution of *Alabama argillacea*

(Hubner) (Lepidoptera: Noctuidae) in cotton. Neotropical Entomology, v. 32, n. 1, p.117-122, 2003.

GALLO, D.; NAKANO, O.; SILVEIRA NETO, S.; CARVALHO, R.P.L.; BATISTA, G.C.; BERTI FILHO, E.; PARRA, J.R.P.; ZUCCHI, R.A.; ALVES, S.B.; VENDRAMIM, J.D.; MACHINI, L.C.; LOPES, J.R.S.; OMOTO, C. Manual de entomologia agrícola. v. 10. Piracicaba: FEALQ, 2002. 920p.

GERBER, K.; GIBLIN, R. Association of the red ring nematode and other nematodes species with the palm weevil Rhynchophorus palmarum L. Journal Nematology, v.22, n.2 p.143- 149, 1990.

GOMES, S. M. S. Evaluation of Capture Systems for Metamasius hemipterus and Rhynchophorus palmarum (Curculionidae) in Plantations of Pupunha (Bactris gasipaes) and Dendê (Elaeis guineensis) in Southern Bahia. 2008. 54 f. Dissertation (Master's Degree in Entomology) - Federal University of Viçosa, Viçosa-MG.

GREGO, C. R.; VIEIRA, S. R.; LOURENÇÃO, A. L. Spatial distribution of Pseudaletia sequax franclemlont in triticale under no-till management. Scientia Agricola, v.63, n.4, p.321-327, 2006.

HASSEL, M.P. Detecting regulation in a patchly distributed animal populations. Journal of Animal Ecology, v.56, p.705-713, 1987.

ISSAKS, E.H.; SRIVASTAVA, R.M. An introduction to applied geostatistics. Oxford: University Press, 1989.

LACERDA, I. J. Coconut trees under attack. Cultivar Hortaliças e Frutas magazine, v.6, n.10, p.25-28, 2001.

LEAL, R. M.; BARBOSA, J. C.; COSTA, M. G.; BELASQUE JUNIOR, J.; YAMAMOTO, P. T.; DRAGONE, J. Spatial distribution of huanglongbing (greening) in citrus using geostatistics. Revista Brasileira de Fruticultura, v. 32, n.3, p. 808-818, 2010.

LEÓN-BRITO, O.; VÁSQUEZ, L. N.; LÁREZ, C.; SILVA-ACUNA, R. Ciclo de vida y longevidad de Metamasius hemipterus L. (Coleoptera: Curculionidae), uma plaga de la palma aceiteira em el estado Monagas, Venezuela. Bioagro, v.17, n.2, p.115-118, 2005.

LIEBHOLD, A. M.; ROSSI, R. E.; KEMP, W. P. Geostatistic and geographic information system in applied insect ecology. Annual review of Entomology, v.38, n.1, p.303- 327, 1993.

MORAIS, R. P.; RODRIGUES, V. K. M.; VERAS, A. T. de R.; NETA, L. C. B. Physiographic aspects of the landscape in São João da Baliza - RR and its favourability for planting oil palm for biodiesel production. REVISTA GEONORTE, Ed. Especial, V.3, N.4, p. 42-50, 2012.

MOURA, J. I. L.; TOMAS, R.; SGRILLO, R.; DELABIE, J. H. C. Natural Efficiency of Parasitism by Billaea rhynchophorae (Blanchard) (Diptera: Tachinidae) for the control of Rhynchophorus palmarum (L) (Coleoptera: Curculionidae). Neotropical Entomology. V. 35, p. 273-274, 2006.

MURPHY D. J. Future prospects for oil palm in the 21st century: biological and related challenges. European Journal of Lipid Science and Technology, v. 109, p. 296306, 2007.

NATIONAL MUSEUM (Brazil). Elaeis guineenses. Rio de Janeiro, UFRJ, Botanical Garden. Available at: <http://www.museunacional.ufrj.br/hortobotanico/elaeisguineensis.htm> Accessed 13 Aug. 2014.

NAVARRO, D. M. A. F.; MURTA, M. M.; DUARTE, A. G.; LIMA, I. S.; NASCIMENTO, R. R.; SANT'ANA, A. E. G. Practical aspects related to the use of rincoforol, the aggregation pheromone of the coconut palm borer Rhynchophorus palmarum L. (Coleoptera: curculionidae) in the control of coconut palm pests. Analysis of its efficiency in the field. Química Nova, v. 25, p. 32-36, 2002.

O'BRIEN, C. W.; THOMAS, M. C. The species of Metamasius in Florida (Coleoptera: Curculionidae). Florida Dept. Agric. & Consumer Services, Division of Plant Industry. Entomology Circular, n. 330, 1990.

RAMALHO FILHO, A. MOTA, P. E. F. Agro-ecological zoning of oil palm for deforested areas of the Legal Amazon. Rio de Janeiro, Embrapa Solos, p. 44. 2010.

RAMOS, I. G. Zeolites as a device for the prolonged release of rincoforol. 2012. 178f. Thesis (PhD in Analytical Chemistry) - Institute of Chemistry, Federal University of Bahia, Salvador.

RICKLEFS, R. E. 2003. The economy of nature, 5ª ed. Rio de Janeiro, Guanabara/Koogan. 470 p.

ROCHA, R. B. Metamasius spp. Horn (COLEOPTERA: CURCULIONIDAE) IN HELICONIAS (ZINGIBERALES: HELICONIACEAE): monitoring, associated organisms and control tactics with Beauveria bassiana and neem-based insecticides. 2012. 77f. Dissertation (Master's in Plant Production), State University of Santa Cruz.

RUIZ, R.; DEMETRIO, C.; ASSUNÇÃO, R.; LEANDRO, R. Bayesian hierarchical models to study the spatial distribution of coffee berry borer infestation at the local level. Revista Colombiana de Estadística. v. 26, n.1, p. 1-24, 2003.

SALES, T. M. Geostatistical Analysis of the Spatial Distribution of Fatal Yellowing of the Oil Palm in the Northeast of Pará. 2011. 32 f. Dissertation (Master's Degree in Agronomy) - Federal Rural University of Amazonia (UFRA).

SÁNCHEZ-SOTO, S.; NAKANO, O. Record of Rhynchophorus palmarum L. (Coleoptera:

Curculionidae) in the State of Mato Grosso do Sul. Neotropical Entomology, v. 31, p. 659-660, 2002.

SANTOS, A. M. Analysing the potential of palm biodiesel for generating electricity in isolated systems in the Amazon. 2008. 224 p. Master's dissertation, Alberto Luiz Coimbra Institute for Postgraduate Studies and Engineering Research/Federal University of Rio de Janeiro.

SECRETARIAT OF AGRICULTURE OF PARÁ - SAGRI. Outstanding crops in the state of Pará-Year 2013. Available at:< http:// www.sagri.pa.gov.br>. Accessed on: 20 Feb. 2013.

SILVA JUNIOR, R. L. da. Spatial variability of the cone index correlated with maps of productivity. 2001. 132p. Dissertation (Master's Degree) - "Luiz de Queiroz" College of Agriculture, University of São Paulo.

SILVA, A. G. da.; FARIAS, P. R. S.; BOIAÇU JUNHO, A. L.; CORREIA, R. G.; SILVA, J. B. da.; RODRIGUES, N. E. L. Spatial analysis of the citrus black fly in a citrus orchard using geostatistics. Revista Agricultura, v.86, n.2, p.102-114, 2011.

SOLIMAN, E. P.; PAVARINI, R.; GARCIA, V. A.; LIMA, R. C.; NOMURA, E. S.; PAVARINI, G. M. P.; DAMATTO JUNIOR, E. R. Different baits for population monitoring of Metamasius sp. (Coleoptera: Curculionidae) in the cultivation of the pupunheira tree. Revista Eletrônica de Agronomia, v. 16, n. 2, p. 1-6, 2009.

STECCA, C. S. Spatio-temporal distribution and population fluctuation of soya bean defoliator caterpillars. 2011. 86 f. Dissertation (Master's Degree in Agronomy) - Federal University of Santa Maria, Santa Maria.

TAN, K.; LEE, K.; MOHAMED, A.; BHATIA, S. Palm oil: Addressing issues and towards sustainable development. Renewable and Sustainable. Energy Reviews, v.13 n.2, p. 420-427. 2009.

TAYLOR, L.R. Assessing and interpreting the spatial distributions of insect populations. Annual Review Entomology, v. 29, p. 321-57, 1984.

TINÔCO, R. S. Natural enemies and defoliator lepidopterans associated with Elaeis guianensis Jacq., in agropalma, Brazilian Amazon. 2008. 51 f. Master's thesis - Federal University of Viçosa.

TRINDADE, D. R.; POLTRONIERI, L. S.; FURLAN, J. Approach to the current state of research into the identification of the causal agent of the fatal yellowing of the oil palm. In: POLTRONIERI, L.S.; TRINDADE, D. R.; SANTOS, I. P. Pests and diseases of Amazonian crops (Ed.). Belém: Embrapa Amazônia Oriental, CPATU, p. 439-450, 2005.

TRIPLEHORN, C. A.; JONNSON, N. F. Study of Insects. São Paulo: Cengage Learning, 809 p.,

2011.

USDA - UNITED STATES DEPARTMENT OF AGRICULTURE. Oil seeds: World markets and Trade. Foreign Agricultural Service. Biofuels, p. 1-11, Dec. 2011 (Circular series). Available at: <http://www.nal.usda.gov>. Accessed on: 16 Dec. 2014.

USDA - UNITED STATES DEPARTMENT OF AGRICULTURE. Oil seeds: World markets and Trade. Foreign Agricultural Service. Biofuels, p. 1-34, Dec. 2013 (Circular series). Available at: <http://www.nal.usda.gov>. Accessed on: 18 October 2014.

VIANNA, F. C. Eco-efficiency Analysis: Evaluation of the Economic-Environmental Performance of Biodiesel and PetroDiesel. 2006. 182 f. Dissertation (Master's in Chemical Engineering), Polytechnic School, University of São Paulo.

VIEIRA, S.R. Geostatistics applied to precision agriculture. In: BORÉN, A. Agricultura de precisão. Viçosa: UFV, 2000, p.93-108.

VIEIRA, S.R.; HATFIELD, J.L.; NIELSEN, D.R.; BIGGAR, J.W. Geostatistical theory and application to variability of some agronomic properties. Hilgardia, v.51, n.3, p.1- 75, 1983.

YASUDA, M. E. Survey of potential pests associated with the cultivation of the Australian royal palm Archontophoenix spp. (Arecaceae) in the Middle Itajaí Valley. 2005. 94f. Dissertation (Master's in Environmental Engineering) - Regional University of Blumenau - FURB, Blumenau-SC.

YOUNG, L. J. and YOUNG, J. H. 1998. Statistical ecology: a population perspective. Boston, Kluwer Academic Publishers. P. 565.

ZORZENON, F.J.; BERGMANM, E.C.; BICUDO, J.E.A. First Occurrence of Metamasius hemipterus (LINNAEUS, 1758) and Metamasius ensirotris (GERMAN, 1824) (Coleoptera, Curculionidae) on Palm Trees of the Genera Euterpe and Bactris (Arecaceae) in Brazil. Archives of the Biological Institute, v. 67, n. 2, p. 265-268, 2000.

CHAPTER 2

SPACE DISTRIBUTION OF Rhynchophorus palmarum L., IN A PLANTATION OF OIL PALM (Elaeis guineensis Jacq) IN THE SOUTH OF THE STATE OF RORAIMA

5.1 SUMMARY

The oil palm (Elaeis guineensis Jacq.) is a palm of African origin that stands out for its high production per unit area, but one of the main obstacles is phytosanitary problems. The aim of this study was to assess the spatial distribution pattern of Rhynchophorus palmarum in oil palm plantations in the south of the state of Roraima. The experimental area is located in the municipality of São João da Baliza, RR, where 24 samples were taken from September 2013 to August 2014. The perimeter of the area was demarcated with GPs, and the sampling grids were 45x45m. To collect the insects, a bucket trap with a food attractant (sugar cane) was placed at each sampling point. Spatial variability and spatial dependence were analysed by incorporating geostatistical procedures based on spatial modelling techniques using semivariograms. Kriging maps were generated from R. palmarum adult count data in the field. The sampling grids proved appropriate for characterising the spatial distribution of R. palmarum in the field. The spatial distribution of R. palmarum is aggregated with spatial dependence described by the spherical model, forming "plots" of 100 to 210 metres in radius. It was observed that *R. palmarum* infestation occurs both in the centre and at the edges of the plantation, with subsequent spread throughout the area. The moderate spatial dependence index prevailed in 23 evaluations, and only one had strong dependence. The maximum range found in the study was 210 metres with an area of influence of 13.9 ha and the minimum was 100 with an area of influence of 3.1 ha.

Keywords: Geostatistics. Integrated pest management. Oil palm.
Semivariogram. Kriging.

5.3 INTRODUCTION

The oil palm (Elaeis guineensis Jacq.) is a species of great agro-ecological-industrial importance for the Amazon, contributing to the development of the region by generating jobs and income (SANTOS 2008; CORDEIRO et al., 2009), due to its high profitability, keeping people in the countryside, as well as being a source of one of the main vegetable oils (ABDALLA et al., 2008). The crop has a commercial exploitation period of approximately 25 years, and if properly managed, plantations produce an average of 4 to 6 tonnes of oil/ha/year (CHIA et al., 2009).

Oil palm cultivation is technologically well-developed and the species is the most productive oilseed and one of the best options for bioenergy production in the agro-ecological conditions of the Amazon, since the plantation is exploited for decades without the need for soil preparation, providing permanent cover and avoiding the direct impact of heavy rains that cause erosion and leaching (CHIA et al., 2009).

Phytosanitary problems are one of the limiting factors in the exploitation of palm cultivation, as they are of significant economic importance (DUARTE et al. 2008). Considered one of the main pests of palm trees throughout tropical America, Rhynchophorus palmarum (L.) (Coleoptera: Curculionidae) can cause direct damage through its larvae and adults. When they hatch, the larvae make galleries in the plant's tissues, mainly in the apical bud, the petiole of new leaves, the inflorescences and the soft stipe, destroying the sap-conducting tissues, which can cause the plant to die (DUARTE et al., 2003; MOURA et al., 2006; AMBROGI et al., 2009). This species is the vector responsible for dispersing the nematode Bursaphelenchus cocophilus (Cobb.), the causal agent of red ring rot (CHINCHILLA 1992). It is worth noting that in a seed production field, this pest can cause irreversible damage, whether from an economic or genetic point of view (CYSNE et al., 2013).

A number of studies have been carried out in recent years dealing mainly with methods of capturing R. palmarum (MOURA et al., 2006; GOMES, 2008; COSTA-CARVALHO et al., 2011; MIGUENS et al., 2011; CYSNE et al., 2013), but specific studies on the spatial distribution of the pest using geostatistics are scarce.

Many studies have been carried out using dispersion indices and geostatistics in order to verify the spatial behaviour of economically important pests in recent years (FARIAS et al., 2004; GREGO et al., 2006; SILVA et al., 2011; MONTEIRO et al., 2013; RIOS et al., 2014; PINHO, 2015), and in most of these studies it has been observed that there is a tendency for insect populations to be distributed in aggregate.

According to Hall and Branham (2008) and Ray et al. (2009), in certain ecological niches it is common for populations of phytophagous insects to distribute themselves in aggregate form. This type of behaviour can bring benefits throughout an insect's life history, such as: encounters between sexual partners during reproduction, defence against natural enemies and greater exploitation of resources (HUNTER, 2000; COSTER-LONGMAN et al., 2002).

Little is known about the spatial distribution of R. palmarum in Amazonian conditions. Given the importance of this pest and the lack of basic studies to implement its proper management in field conditions, the aim of this study was to determine the spatial distribution of R. palmarum in oil palm plantations in the south of the state of Roraima, using geostatistical analysis, in order to guide sampling procedures in the field.

5.4 MATERIAL AND METHODS

5.4.1 Characterisation of the region of the experiment

The study was carried out in a 25 ha area of commercial oil palm (Elaeis guineensis) plantations, 5 years old, in the municipality of São João da Baliza, located at coordinates: latitude 28° 31' 05.53" W and longitude 50° 52' 50.54" N, in the south of the state of Roraima (Figure 1).

The oil palm plants are arranged in the shape of an equilateral triangle with a spacing of 9x9 metres between plants and 7.8 metres between rows. The perimeter of the area was demarcated using a Garmin GPS model GPSmap76CSx. The experimental area was then divided into regular sampling grids. The size of the sampling grid was 45 x 45 m, making up 100 sampling units made up of around 25 plants each, totalling 2,500 plants in the useful area (Figures 1 and 2).

Figura 1. Experimental area in the municipality of São João da Baliza, RR, November 2014. Photo: Luiz Fernandes Silva Dionisio

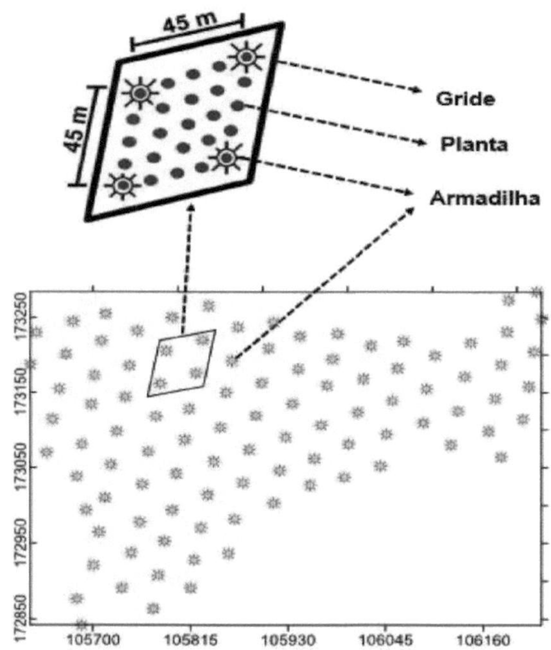

Figura 2. Location of the 100 traps distributed in a 5-year-old commercial oil palm plantation (Elaeis guineensis Jacq) and details of the sampling grid (São João da Baliza - RR).

The region has flat to undulating terrain, with predominantly Submontane Dense Ombrophilous Forest vegetation (IBGE, 2012). According to the Koppen classification, the climate of the southern region of the state of Roraima is of the Am type, characterised as tropical rainy with an average temperature between 25 and 28 °C and annual rainfall ranging from 1,800 to 1,900 mm. The relative humidity is considered high, with the annual average oscillating between 85 and 90 per cent, and the luminosity in the region varies from 1,500 to 3,000 hours/year of solar radiation (BARBOSA, 1997).

5.4.2 Setting up the traps

Bucket traps containing the food attractant sugar cane were used for mass collection of R. palmarum. These traps were made up of 15-litre plastic buckets, the bottom of which was perforated to facilitate the drainage of rainwater. A hole was drilled in the lid of the bucket and a 10 cm diameter plastic funnel was placed in its ventral part without the narrow tube, which was removed to allow the insects to pass into the bucket (Figure 3).

Figure 3 - Bucket traps (A) - preparing the sugar cane; (B) - assembling the trap; (C) - trap ready to catch insects; (D) - view of the hole in the top of the lid. Photo: Luiz Fernandes Silva Dionisio

Inside each trap, 15 sugar cane stalks (food attractant) 20 cm long were placed and crushed to facilitate the volatilisation of the odour, with the aim of attracting the insects.

Every fortnight, from September 2013 to August 2014, the R. palmarum individuals were removed from the traps and placed in jars with 70% alcohol. On this occasion, the sugar cane stalks were also replaced.

5.4.3 Analysing data

After tabulating the data, the spatial distribution of R. palmarum was studied using geostatistics, using the semivariogram, adjusting one of the 4 possible models, Gaussian, spherical, power or exponential.

The first stage of the geostatistical study was to obtain the semivariograms, which were adjusted to a model that provided the radius of aggregation, and from these semivariograms, maps of the organism's expansion in the area were drawn up (LEAL et al., 2010). The semivariogram is estimated using the mathematical expression:

$$\gamma(h) = \frac{1}{2\,N\,(h)} \sum_{i=l}^{N(h)} [Z(x_i) - Z(x_i + h)]^2$$

where,

- $\gamma(h)$*is* the semivariance
- *N(h)* is the number of observation pairs [*Z*(*xi*); Z(*xi*+*h*)] separated by distance h.

When calculating the semivariogram, pairs of semivariogram values [(*y hhé*] and distances (*h*) were obtained and arranged on a scatter plot, with the *y* values being the semivariance and the x values being the distances. A model was fitted to these points.

To better define the variogram parameters, several attempts were made to fit the spherical and exponential models, using a standard distance (range) of 240 m to model the semivariograms. All the adjustments to estimate the spatial dependence of the variables used a number of pairs greater than 30, as recommended by Guerra (1988). As for the maximum distance, the same author mentions that in practice the variographic study area rarely exceeds half of the total field sampled.

In this study, the semivariogram model that best fitted the data was the spherical model, which is described as:

$$\gamma(h) = C_0 + C_1\left[\frac{2}{3}\left(\frac{h}{a}\right) - \frac{1}{2}\left(\frac{h}{a}\right)^3\right] \rightarrow 0 < h < a$$

$$\gamma(h) = C_0 + C_1 \rightarrow h \geq a$$

- Co is the nugget effect or minimum semivariance;
- Co+ Ci is the plateau or maximum semivariance;

- a is the range or radius of aggregation.

All the analyses followed the procedures described by Vieira et al. (1983), who used information on the position of the sample and the value that the variable assumed at each point. In this study, the assumed variable was (incidence of R. palmarum) and the point is represented by the trap. In this way, the value of the variable and the coordinates (latitude and longitude) were obtained from each sampling point and used to construct R. palmarum incidence maps using the kriging interpolation method (SILVA et al., 2011).

The parameters relating to the semivariogram were defined: the plateau (C0 + C1), the nugget effect (C0) and the range (a). Where the plateau is the value of the semivariogram corresponding to its range, the nugget effect is the semivariance at very close points, when the distances between the sampling units are very small and the range (a) represents the distance at which there is spatial dependence between the samples, and its value has been used to determine the appropriate distance between the sampling points (SHAO-KUI et al., 2005; SILVA et al., 2011).

The Spatial Dependence Index k (ratio between C0/(C0+C1)) was also calculated to determine the spatial dependence of each assessment according to Cambardella et al. (1994), who classify spatial dependence as weak (SDI > 75 %), moderate (25 % < SDI < 75 %) and strong (SDI < 25 %).

Both the geostatistical analysis and the population maps were drawn up using the statistical programme SURFER Version 8.0 (2002)

5.5 RESULTS AND DISCUSSION

Spatial dependence was observed among the data analysed, and the distribution of R. palmarum occurred in aggregate form in all the evaluations. The spherical model was the best fit in the semivariograms, and based on its conformation the spatial dependence of the pest was verified, indicating the formation of clusters of 100 to 210 m in radius (Figures 4 and 5). The aggregate distribution established by the spherical model has been the most common in insect studies (FARIAS et al., 2004), and is characterised by the formation of "clusters" in the field where these insects group together (SILVA et al., 2011).

Figures 4 and 5 show the semivariograms of the spatial distribution of R. palmarum. The coefficient of determination (R^2) indicates the quality of the fit of the semivariogram model. Values close to 1 (one) indicate a good model fit (SILVA et al., 2011), which was observed in this study, where this parameter ranged from 0.91 to 0.99.

The range of spatial dependence is of great importance because, as well as indicating the radius of aggregation of R. palmarum, it allows us to recommend the radius of eradication of these insects. Using the radius of aggregation, it can be said that, under the conditions of the study, the radius of eradication of R. palmarum ranged from 100 to 210 metres.

The maximum range found in this study was 210 metres with an area of influence of 13.9 ha and the minimum was 100 metres with an area of influence of 3.1 ha. The use of the sampling grid (45x45m) proved to be adequate and made it possible to correctly detect the spatial distribution of R. palmarum, as the minimum range found was 100 metres. With these results, it can be said that in order to obtain a reliable estimate of the pest population, a trap is needed every 13.9 ha, starting from the highest range value (210 m).

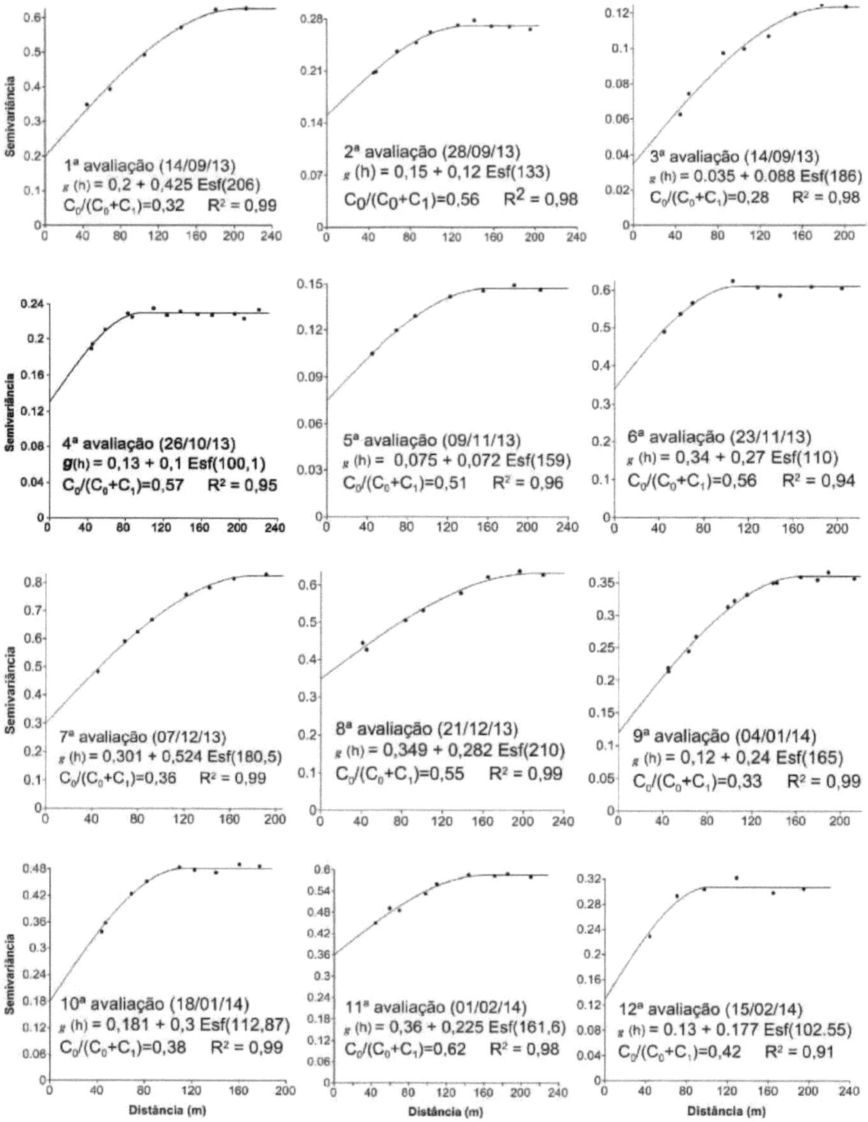

Figure 4: Semivariograms of the spatial distribution of R. palmarum in the experimental area. São João da Baliza, RR, from September 2013 to February 2014.

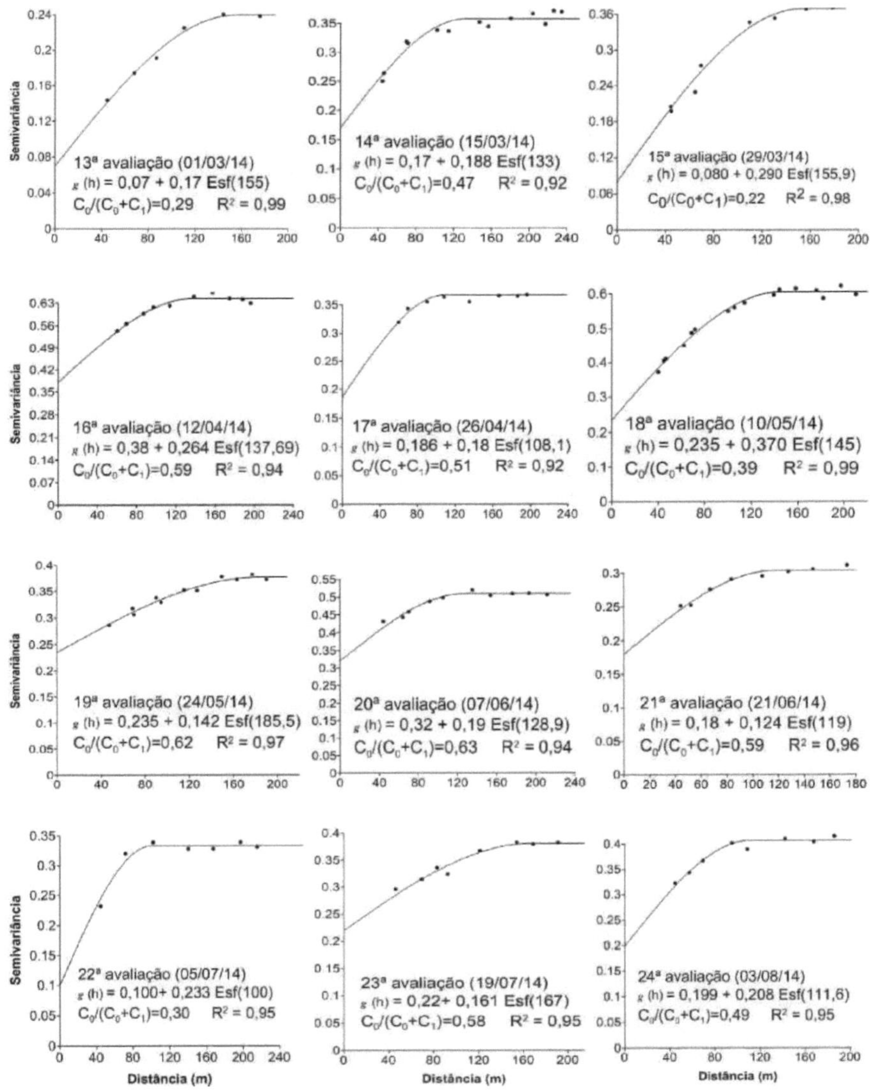

Figura 5. Semivariograms of the spatial distribution of R. palmarum in the experimental area. São João da Baliza, RR from Mar/2014 to Aug/2014.

Dal Prá et al. (2011), when studying the spatial distribution of Diloboderus abderus Sturm, 1826 (Coleoptera: Melolonthidae), found a range that varied from 125 to 195 m. Fernandes (2009), working with Hypothenemus hampei Ferrari, 1867 (Coleoptera: Scolitydae) found a range of 55 to 375 m, while Pinho (2015), assessing the spatial distribution of Rhynchophorus palmarum Linnaeus, 1758 (Coleoptera: Curculionidae) in oil palm agroecosystems using pheromone in the state of Pará, found a range of 175 to 710 m.

The aggregated distribution found in this study for R. palmarum was also evidenced by Faleiro et al. (2010) in Saudi Arabia, who using mathematical frequency distribution models concluded that the population of R. ferrugiensis Olivier, 1790 (Coleoptera: Curculionidae) follows an aggregated pattern. Pinho (2015), assessing the spatial distribution of R. palmarum Linnaeus, 1758 (Coleoptera: Curculionidae) using geostatistics in an oil palm plantation in the state of Pará, also found that this insect is distributed in an aggregated manner.

The C0/(C0+C1) ratio provides a measure to estimate the degree of dependence in the samples (clustering), i.e. how much randomness there is in the surveys. The values observed in this study ranged from 0.22 to 0.63 between neighbouring points on the various sampling dates (Figures 4 and 5). These values are within the range recommended by Journel and Hijbregts (1978), who state that values below 0.80 indicate an aggregated distribution of the variable and values greater than 0.80 indicate that the phenomenon studied is tending towards randomness and that there is no spatial dependence between the samples.

Spatial dependence was classified as strong in only one of the 24 evaluations, according to the classification of Cambardella et al. (1994), because in this sample the value of the nugget effect was less than 25% of the value of the plateau. In the other samples, spatial dependence was classified as moderate because the value of the nugget effect was greater than 25% and less than 63% of the value of the plateau.

To better visualise the spatial distribution of R. palmarum, kriging maps were drawn up. It was initially observed that there were some outbreaks of R. palmarum and, later, the formation of plagues of the pest, which expanded towards the edges of the plantation. However, R. palmarum tends to change its infestation site over time. In some evaluations, the presence of this individual could be seen both at the edges and in the centre of the plantation. The size of these plots was determined according to when there was the highest or lowest incidence of R. palmarum adults (Figures 6 to 9).

Thus, in the evaluations that showed the lowest incidence of R. palmarum adults, larger plots were formed. This occurred in the following samples: 1ª to 5[(a)], 13ª to 20[(a)], 22[(a)]to 24[(a)]) (Figures 6, 8 and 9), which correspond to the following evaluation periods: September/2013 to November/2013 and March/2014 to August/2014.

However, when the incidence of R. palmarum in the area was higher, several smaller plots of the pest formed, as can be seen in the samples: 6th to 12th and 21st (Figures 6 and 7). The highest population densities of R. palmarum occurred from November 2013 to February 2014, months that correspond to the dry season. It was also found that the highest densities in the aforementioned months coincided with the period in which the crop underwent cultural treatments (mechanised weeding, cutting of straw and application of herbicide (Glyphosate). Correia (2012), studying the population fluctuation of R. palmarum in oil palm plantations in the south of the state of Roraima, found that this insect showed higher population peaks in the months of March, September, December and January.

Knowing the spatio-temporal dynamics of insects in agricultural ecosystems is important for developing management strategies and reducing the use of pesticides (BLACKSHAW; VERNON, 2006), providing better informed decision-making due to the greater amount of information obtained (DAL PRÁ et al., 2011). Therefore, maps showing the occurrence and distribution of insect pests in the area can contribute to more effective sampling and control through the application of phytosanitary products in the necessary quantities.

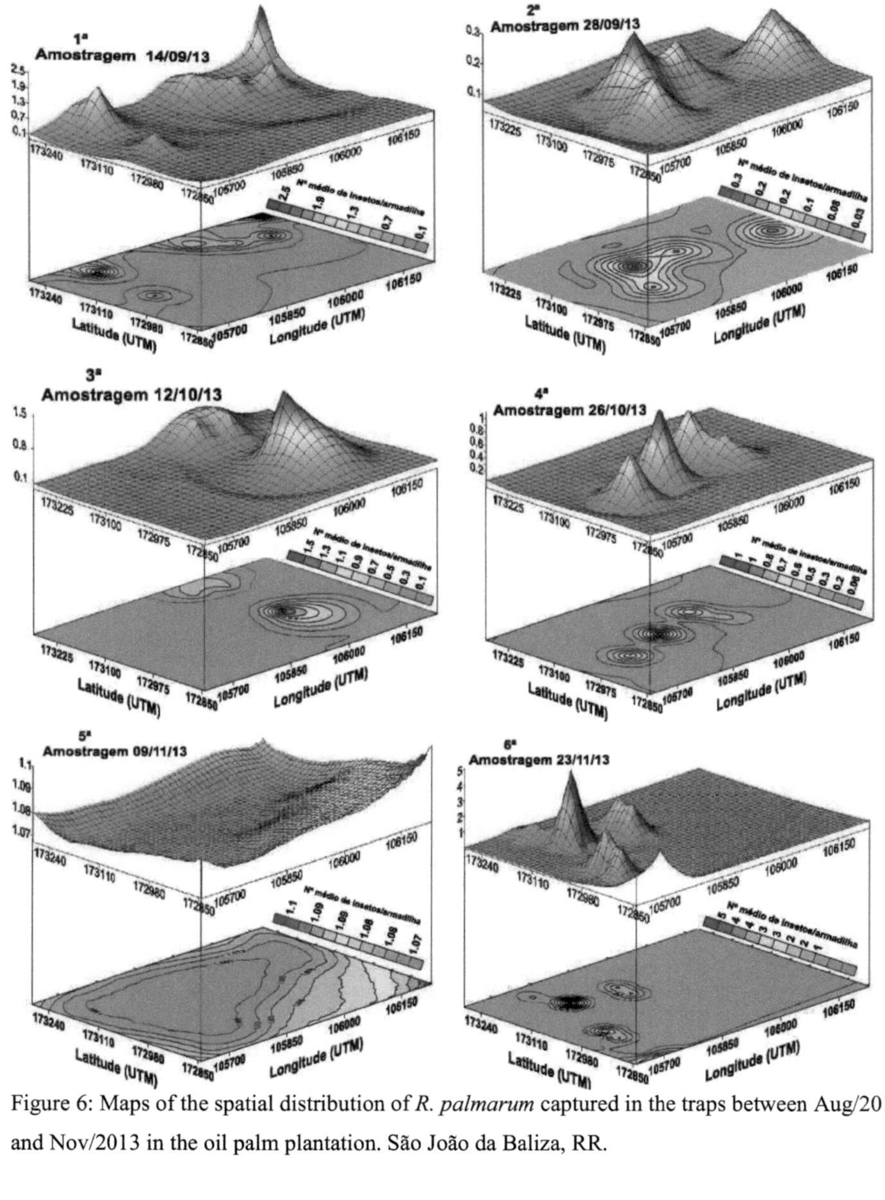

Figure 6: Maps of the spatial distribution of *R. palmarum* captured in the traps between Aug/2013 and Nov/2013 in the oil palm plantation. São João da Baliza, RR.

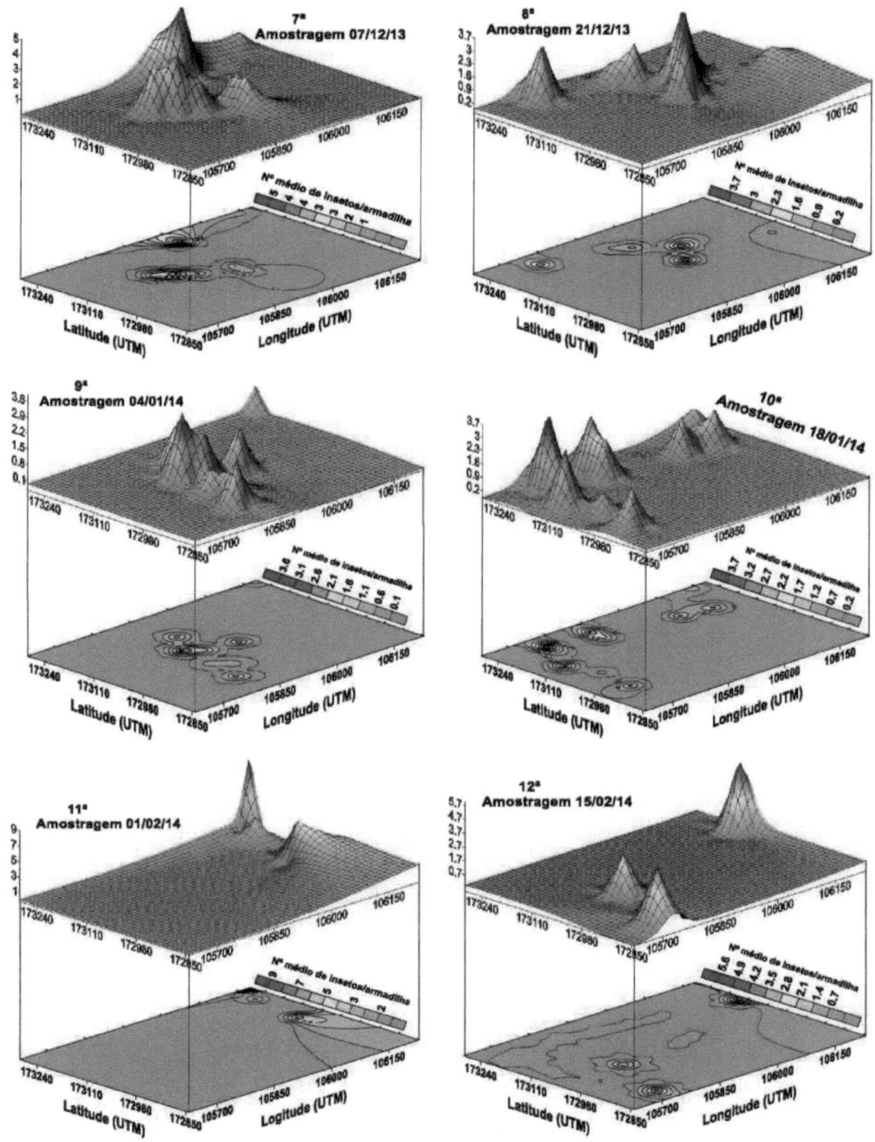

Figura 7. Maps of the spatial distribution of *R. palmarum*, captured in the traps, from Dec 2013 to Feb 2014 in a palm oil plantation. São João da Baliza, RR.

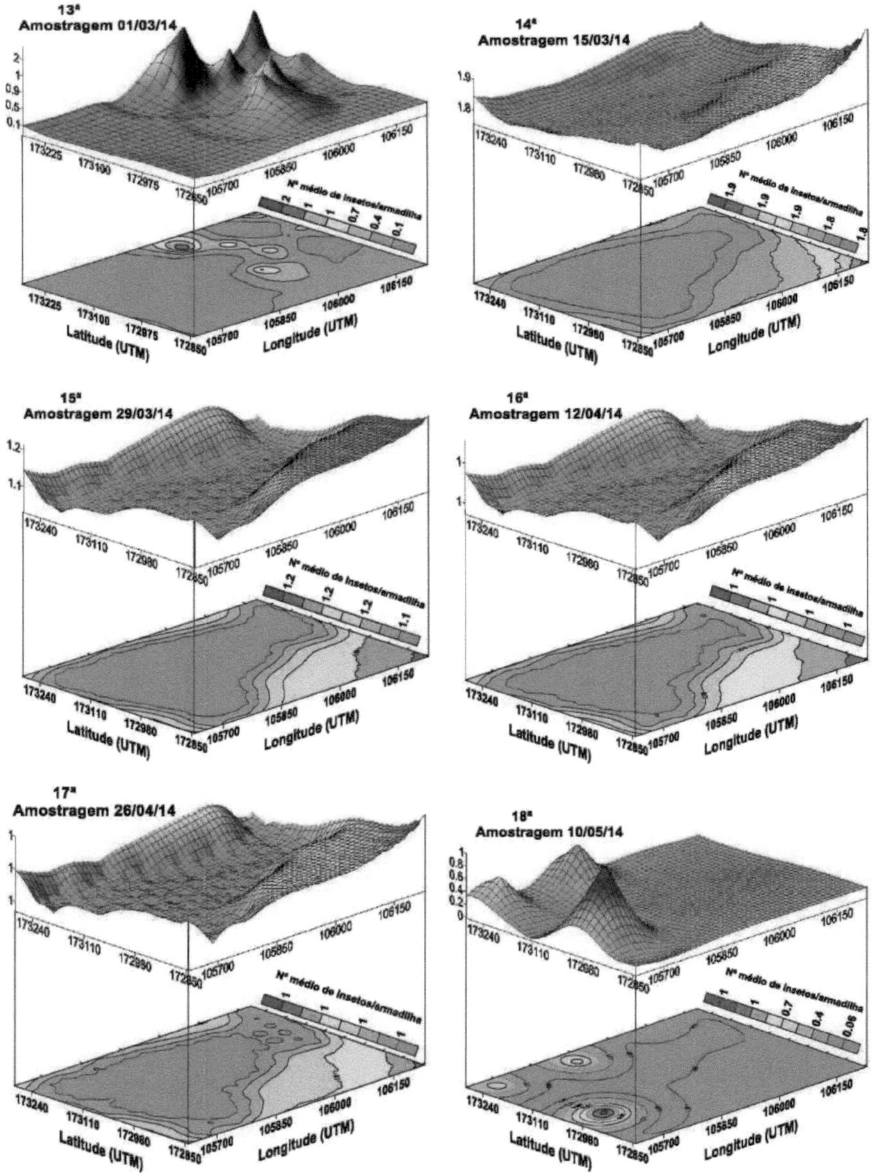

Figura 8. Maps of the spatial distribution of *R. palmarum*, captured in the traps, from March 2014 to May 2014 in oil palm plantations. São João da Baliza, RR.

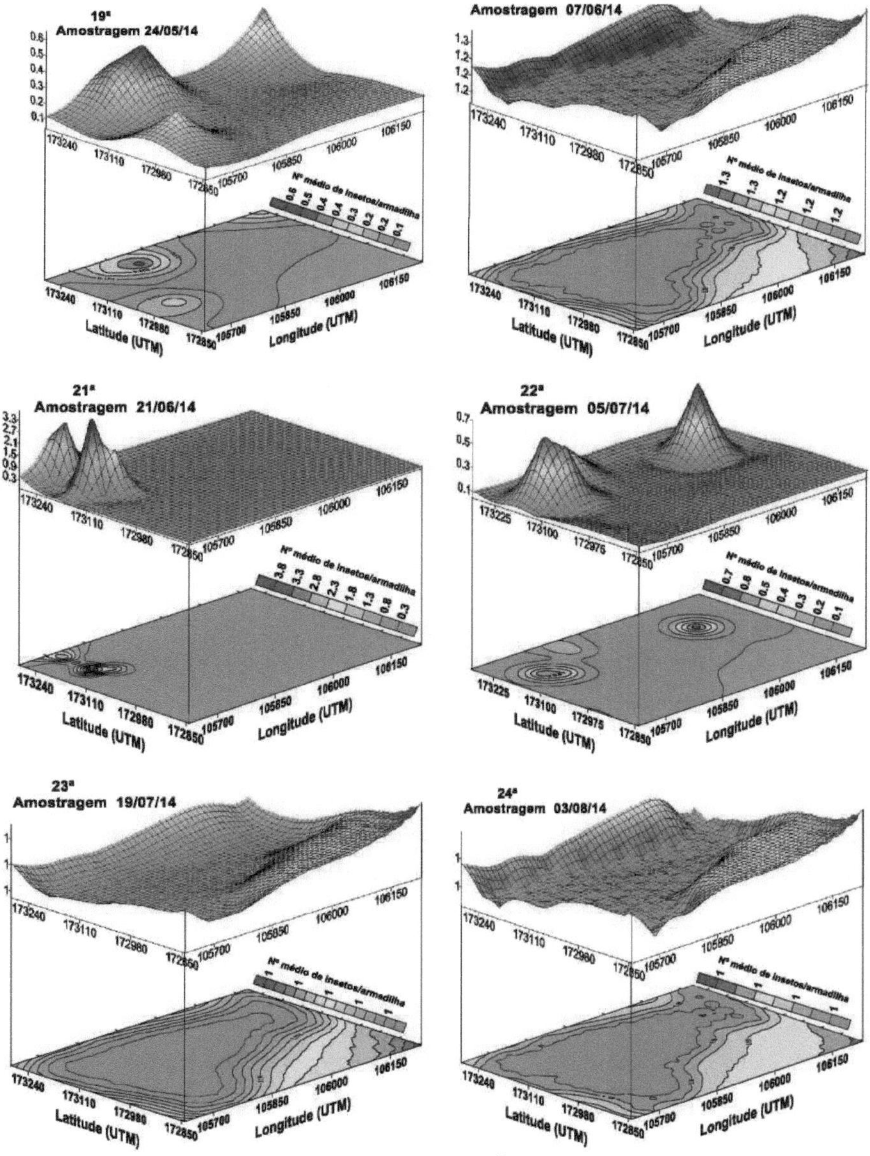

Figura 9. Maps of the spatial distribution of *R. palmarum*, captured in the traps, from May/2013 to Aug/2014 in oil palm plantations. São João da Baliza, RR.

5.6 CONCLUSIONS

The spatial distribution of R. palmarum in the study area was aggregated, with spatial dependence described by the spherical model, forming "plots" of 100 to 210 metres in radius.

R. palmarum infestation occurs both in the centre and at the edges of the plantation, with subsequent spread throughout the area.

In order to develop safe R. palmarum sampling methods, it is recommended that one trap be distributed every 13.9 ha, using a food attractant.

5.7 BIBLIOGRAPHICAL REFERENCES

ABDALLA, A. L.; SILVA FILHO, J. C.; GODÓI, A. R.; CARMO C. A.; EDUARDO, J. L. P. Utilisation of by-products from the biodiesel industry in ruminant feed. Revista Brasileira de Zootecnia. v.37, n. Esp, p. 258-260. 2008.

AMBROGI, B. G.; VIDAL, D. M.; ZARBIN, P. H. G.; ROSADO-NETO, G. H. Aggregation pheromones in curculionidae (Insecta: Coleoptera) and their taxonomic implication. Química Nova, v. 32, n. 8, p. 2151-2158, 2009.

BARBOSA, R. I. Rainfall distribution in Roraima. In: INPA. Man, Environment and Ecology in the State of Roraima. Manaus, 1997. p. 325-335.

BASOALTO, E.; MIRANDA, M.; KNIGHT, A. L.; FUENTES-CONTRERAS, E. Landscape analysis of adult codling moth (Lapidoptera: Tortricidae) distribution and dispersal within typical agroecosystems dominated by apple production in central Chile. Environmental Entomology, v. 39, n. 5, p. 1399-1408, 2010.

BLACKSHAW, R.P.; VERNON, R.S. Spatio temporal stability of two beetle populations in non-farmed habitats in an agricultural landscape. Journal of Applied Ecology, v.43, n.4, p.680-689, 2006.

CAMBARDELLA, C.A.; MOORMAN, T. B.; PARKIN, T. B.; KARLEN, D. L.; NOVAK, J. M.; TURCO, R. F.; KONOPKA, A. E. Field scale variability of soil properties in central Iowa soils. Soil Science Society of America Journal, v.58, n.6, p.1501- 1511, 1994.

CHIA, G. S.; LOPES, R.; CUNHA, R. N. V.; ROCHA R. N. C.; LOPES M. T. G. Repeatability of bunch production of interspecific hybrids between kayak and oil palm. Acta Amazonica, v. 39, n.2, p. 249-254, 2009.

CHINCHILLA, C.M. El síndrome Del anillo-rojo hoja pequena em palma aceitera e cocotero. Revista Palms, v. 13, n. 1, p. 33-55, 1992.

CORDEIRO, A. C. C.; MACIEL, F. C. da S.; SMIDERLE, O. J.; ALVES, A. B. Vegetative development of oil palm in savannah and forest ecosystems in Roraima. Boa Vista, Embrapa

Roraima. Research and Development Bulletin, n.10, p. 16, 2009.

CORREIA, R. G. POPULATION FLUCTUATION OF Rhynchophorus palmarum, Metamasius hemipterus, AND THE NEMATOID Bursaphelenchus cocophilus IN GENDER (Elaeis guineensis) PLANTATIONS IN RORAIMA. 2012, 67 f. Dissertation (master's degree in agronomy). Federal University of Roraima.

COSTA-CARVALHO, R. R. da.; WARWICK, D. R. N.; SOUZA, P. E. de.; CARVALHO FILHO, J. L. S. de. Longevity of Thielaviopsis paradoxa, causal agent of coconut resinosis in Rhynchophorus palmarum. Scientia Plena, v. 7, n. 4, p. 1-6, 2011.

COSTER-LONGMAN, C.; LANDI, M.; TURILLAZZI, S. The role of passive defence (Selfish Herd and Dilution Effect) in the gregarious nesting of Liostenogaster wasps (Vespidae, Hymenoptera, Stenogastrinae). Journal of Insect Behavior, v.15, n.3, p.331-350, 2002.

CYSNE, A. Q.; CRUZ, B. A.; CUNHA, R. N. V. da.; ROCHA, R. N. C. da. Population fluctuation of Rhynchophorus palmarum (L.) (Coleoptera: Curculionidae) on oil palms in Amazonas. Acta Amazonica. v. 43, n.2, p. 197-202, 2013.

DAL PRÁ, E.; GUEDES, J. V. C.; CHERMAN, M. A.; JUNG, A. H.; SILVA, S. J. P. da.; RIBAS, G. G. Use of geostatistics to characterise the spatial distribution of Diloboderus abderus larvae. Ciência Rural, v.41, n.10, p. 1689-1694, 2011.

DINARDO-MIRANDA, L. L.; VASCONCELOS, A. C. M.; VIEIRA, S. R.; FRACASSO, J. V.; GREGO, C. R. Use of geostatistics in the evaluation of the spatial distribution of Mahanarva fimbriolata in sugarcane. Bragantina, v.66, n. 3, p. 449-455, 2007.

DUARTE, A. G.; LIMA, I. S.; NAVARRO, D. M. A. F.; SANT'ANA, A. E. G. Capture of Rhynchophorus palmarum L. (Coleoptera: Curculionidae) in traps baited with the aggregation pheromone and volatile compounds from pineapple fruit. Revista Brasileira de Fruticultura, v. 25, n. 1, p. 81-84, 2003.

DUARTE, A.G.; LIMA, I.S.; ARAÚJO JÚNIOR, J.V.; DUARTE, A.G.; ALBUQUERQUE, A.L.S.; CRUZ, M.M. Disposition of the nematode Bursaphelenchus coccophilus (Cobb) Baujard, in coconut trees with red ring disease. Revista Brasileira de Fruticultura, v. 30, n. 3, p. 622-627, 2008.

FALEIRO, J. R.; ABDALLAH BEN ABDALLAH.; ASHOK KUMAR, J.; ABDALLAH SHAGAGH.; SAAD AL ABDAN. Sequential sampling plan for area-wide management of Rhynchophorus ferrugineus (Olivier) in date palm plantations of Saudi Arabia. International Journal of Tropical Insect Science, v. 30, n. 3, p. 145-153, 2010.

FARIAS, P. R.S.; BARBOSA, J. C.; BUSOLI, A. C.; OVERAL, W. L.; MIRANDA, V. S.; RIBEIRO, S. M. Spatial analysis of the distribution of Spodoptera frugiperda (J.E. Smith) (Lepidoptera: Noctuidae) and losses in maize crop productivity using geostatistics. Neotropical Entomology, v.37, n.3, p.321-327, 2008.

FARIAS, P.R.S.; ROBERTO, S.R.; LOPES, J.R.S.; PERECIN, D. Geostatistical characterisation of the spatial distribution of Xyllela fastidiosa sharpshooter vectors on citrus. Neotropical Entomology, v.33, n.1, p. 13-20, 2004.

FERNANDES, F. L. Spatial distribution and control decision-making programme using traps for Hypothenemus hampei. 2009, 117 f. Thesis (Doctorate in Entomology) - Federal University of Viçosa, Vicosa, MG.

GOMES, S. M. S. Evaluation of Capture Systems for Metamasius hemipterus and Rhynchophorus palmarum (Curculionidae) in Plantations of Pupunha (Bactris gasipaes) and Dendê (Elaeis guineensis) in Southern Bahia. 2008. 54 f. Dissertation (Master's Degree in Entomology) - Federal University of Viçosa, Viçosa-MG.

GREGO, C. R.; VIEIRA, S. R.; LOURENÇÃO, A. L. Spatial distribution of Pseudaletia sequax franclemlont in triticale under no-till management. Scientia Agricola, v.63, n.4, p.321-327, 2006.

GUERRA, P. A. G. 1988. Operational geostatistics. Ministry of Mines and Energy, 145p.

HALL, D.W; BRANHAM, M.A. Aggregation of Calopteron discrepans (Newman) (Coleoptera: Lycidae) larvae prior to pupation. Florida Entomologist, v.91, n.1, p.124- 125, 2008.

HUNTER, A.F. Gregariousness and repellent defences in the survival of phytophagous insects. Oikos, v.91, n.1, p. 213-224, 2000.

BRAZILIAN INSTITUTE OF GEOGRAPHY AND STATISTICS - IBGE. Technical manual of Brazilian vegetation. 2. ed. Rio de Janeiro: 2012. 275p.

JOURNEL, A.G.; HIJBREGTS, C. J. 1978. Mining geostatistics. London: Academic Press, 600p.

KARIMZADEH, R.; HEJAZI, M. J.; HELALI, H.; IRANIPOUR, S.; MOHAMMADI, S. A. Assessing the impact of site-specific on control of Eurygaster integriceps (Hemiptera: Scutellieridae) damage and natural enemies. Precision Agriculture, v. 12, n.4, p. 576-593, 2011.

LEAL, R. M.; BARBOSA, J. C.; COSTA, M. G.; BELASQUE JUNIOR, J.; YAMAMOTO, P. T.; DRAGONE, J. Spatial distribution of huanglongbing (greening) in citrus using geostatistics. Revista Brasileira de Fruticultura, v. 32, n. 3, p. 808-818, 2010.

MIGUENS, F. C.; MAGALHÃES, J. A. S. de.; AMORIM, L. M. de.; GOEBEL, V. R. COUSTOUR, N. L.; LUMMERZHEIM, M.; MOURA, J. I. L.; COSTA, R. M. Mass Trapping and Biological Control of Rhynchophorus palmarum L.: A hypothesis based on morphological evidences. EntomoBrasilis, v.4, n.2, p.49-55, 2011.

MONTEIRO, O. M.; LUNZ, A. M.; AZEVEDO, R. de; OLIVEIRA JÚNIOR, M. C. M.de. Spatial distribution and gallery depths of Quesada gigas nymphs in parica plantations. Revista Ciencias Agrarias, v. 56, n. 4, p. 353-358, 2013.

MOURA, J. I. L.; TOMAS, R.; SGRILLO, R.; DELABIE, J. H. C. Natural Efficiency of Parasitism by Billaea rhynchophorae (Blanchard) (Diptera: Tachinidae) for the control of Rhynchophorus palmarum (L) (Coleoptera: Curculionidae). Neotropical Entomology, v. 35, n. 2, p. 273-274, 2006.

PINHO, R. C. Spatial Distribution of Rhynchophorus palmarum (Linnaeus, 1758) (Coleoptera: Curculionidae) in an Oil Palm Agroecosystem in Eastern Amazonia. 2015. 54 f. Dissertation (Master's in Agronomy) - Federal Rural University of Amazonia.

RAY, A.M.; MILLAR, J.G.; MCELFRESH, J.S.; SWIFT, I.P.; BARBOUR, J.D.; HANKS, L. M. Male-produced aggregation pheromone of the cerambycid beetle Rosalia *funebris*. Journal of Chemical Ecology, v.35, n.1, p.96-103, 2009.

RIOS, E. S.; MARTINS, I. C. F.; NORONHA, M. P. de.; SILVA, J. A. da.; SILVA FILHO, J. G. da.; BADJI, C. A. Spatial distribution of Spodoptera frugiperda in the wasteland of southern Pernambuco state, Brazil. Revista Ciencias Agrarias, v. 57, n. 3, p. 297304, 2014.

SANTOS, A. M. Analysing the potential of palm biodiesel for generating electricity in isolated systems in the Amazon. 2008. 224 f. Master's thesis. Alberto Luiz Coimbra Institute for Postgraduate Studies and Engineering Research/Federal University of Rio de Janeiro.

SHAO-KUI, G. E.; RAYMOND, I. C.; ZU-FEI, M. A.; GUANG-XUE, Z.; DIAN-MO, L. I. Spatial heterogeneity and population risk analysis of cotton bollworm, Helicoverpa armigera, in China. Insect Science, v.12, n.4, p.255-262, 2005.

SILVA, A. G. da.; FARIAS, P. R. S.; BOIAÇU JUNHO, A. L.; CORREIA, R. G.; SILVA, J. B. da.; RODRIGUES, N. E. L. Spatial analysis of the citrus black fly in a citrus orchard using geostatistics. Revista Agricultura, v.86, n.2, p.102-114, 2011.

SURFER version 8. Maps and semivariograms, New York: Golden Software, 2002.

VIEIRA, S.R.; HATFIELD, J.L.; NIELSEN, D.R.; BIGGAR, J.W. Geostatistical theory and application to variability of some agronomic properties. Hilgardia, v.51, n.3, p.1- 75, 1983.

CHAPTER 2

spatial distribution of metamasius hemipterus in an oil palm plantation (elaeis guineensis jacq) in the south of the state of roraima

6.1 SUMMARY

The oil palm (Elaeis guineensis Jacq.) is a palm of African origin, and in Brazil the largest cultivated areas are in the Amazon region. The aim of this study was to assess the spatial distribution pattern of Metamasius hemipterus in an oil palm (Elaeis guineensis) plantation in the state of Roraima. The experimental area is located in the municipality of São João da Baliza, RR, where 24 samples were taken from September 2013 to August 2014. The perimeter of the area was demarcated using a global positioning system receiver, and the sampling grids were 45x45m. Bucket traps with a food attractant (sugar cane) were used to collect the insects. Spatial variability and spatial dependence were analysed by incorporating geostatistical procedures based on spatial modelling techniques using semivariograms. The kriging maps were generated from M. hemipterus adult count data in the field. The sampling grids proved appropriate for characterising the spatial distribution of M. hemipterus in the field. The spatial distribution of M. hemipterus is aggregated with spatial dependence described by the spherical model, forming "plots" of 78 to 199 m (range of the model). The kriging maps showed that *M. hemipterus* infestation occurs predominantly at the edges of the plantation. Moderate spatial dependence prevailed in 20 evaluations, with only two having strong dependence. The maximum range found in the study was 199 metres with an area of influence of 12.4 ha and the minimum was 78 metres with an area of influence of 1.9 ha.

Keywords: Oil palm. Geostatistics. Semivariogram. Kriging.

6.3 INTRODUCTION

The oil palm (Elaeis guineensis Jacq.) is of African origin and in Brazil the largest cultivated areas are in the Amazon region (VIANNA, 2006; BRAZILIO et al., 2012). Considered a source of one of the main vegetable oils (ABDALLA et al., 2008), oil palm stands out for its high profitability, for keeping people in the countryside and for reducing environmental impacts. In addition to its adaptability, oil palm is a good alternative for the recovery and sustainable use of degraded areas, reconciling environmental sustainability with economic and social development (CORDEIRO et al., 2009).

Brazil has been expanding its oil palm production areas, but there are factors that can influence the productivity of this crop, resulting in major losses in some regions. One of these factors is the occurrence of pests and the damage they cause, which limits the exploitation of palm cultivation, causing economic losses for oil palm producers (DUARTE et al., 2008).

Among the pests that attack oil palm, Metamasius hemipterus stands out. Species of the genus Metamasius are found on all continents, mainly in the tropical region of the Americas, and approximately 110 species have been described worldwide (ROCHA, 2012; ZORZENON et al., 2000). Several species of Metamasius are reported as pests of economic importance in the state of

Florida in the United States (SOLIMAN et al., 2009).

In Brazil, Metamasius hemipterus is associated with several hosts: açaizeiro (Euterpe oleracea Martius), coconut (Cocos nucifera), oil palm (Elaeis guineensis), palmito-juçara (Euterpe edulis Martius), pupunheira, tamareira (Phoenix dactylifera), Australian royal palm (Archontophoenix spp.) all belonging to the Arecaceae (Arecales) family, as well as sugar cane (Saccharum officinarum Linnaeus) and banana *(Musa* spp. - Musaceae) (GOMES, 2008; YASUDA, 2005; ZORZENON et al., 2000).

In general, the damage caused by curculionids is caused during the larval stage and by the perforation of tissues for oviposition and/or feeding (AMBROGI et al., 2009). The species *Metamasius hemipterus*, commonly known as the 'striped borer', is considered an important pest of palm trees because its larvae bore into the plant's stem, causing losses due to reduced tillering and the opening of holes that can serve as a gateway to phytopathogens (SOLIMAN et al., 2009). These factors represent an invaluable loss in the case of a seed production field with considerable economic or genetic value (CYSNE et al., 2013).

A number of studies have been carried out in recent years, mainly looking at methods of catching M. hemipterus (LEÓN-BRITO et al., 2005; GOMES, 2008; SOLIMAN et al., 2009; OLIVEIRA et al., 2010; MOLIN; BARRETO, 2012; ROCHA, 2012), but no specific studies have been carried out on the spatial distribution of the pest. Studies into the spatial distribution of pests have become more widespread in recent years, and these have made it possible to control them more adequately (GREGO et al., 2006; MARUYAMA et al., 2006; TOLEDO et al 2006; DINARDO-MIRANDA et al., 2007; FARIAS et al., 2008; NICKELE et al., 2010; SILVA et al., 2011; MONTEIRO et al., 2013; RIOS et al., 2014; PINHO, 2015).

Given the importance of this pest and the lack of basic studies to implement appropriate management in field conditions, the aim of this study was to characterise the spatial distribution of M. hemipterus in oil palm plantations in the state of Roraima, using geostatistical analysis, in order to guide sampling procedures in the field.

6.4 MATERIAL AND METHODS

6.4.1 Characterisation of the region of the experiment

The study was carried out in a 25 ha area of commercial oil palm plantations (Elaeis guineensis Jack), 5 years old, in the municipality of São João da Baliza, located at the coordinates: latitude 28° 31' 05.53" W, longitude 50° 52' 50.54" N, south of the state of Roraima.

The oil palm plants are arranged in the shape of an equilateral triangle with a spacing of 9x9 metres between plants and 7.8 metres between rows. The perimeter of the area was demarcated using a Garmin GPS model GPSmap76CSx and the experimental area was then divided into regular sampling grids. The size of the sampling grid was 45 x 45 m, making up 100 sampling units made up of around 25 plants each, totalling 2,500 plants in the useful area.

The region has flat to undulating terrain, with the predominant vegetation considered to be Submontane Dense Ombrophilous Forest (IBGE, 2012). According to the Koppen classification, the climate of the southern region of the state of Roraima is of the Ami type, characterised as tropical rainy with an average temperature between 25 and 28 °C and annual rainfall ranging from 1,800 to 1,900 mm. The relative humidity of the air is considered high, with the annual average oscillating between 85 and 90 per cent, and the luminosity in the region varies from 1,500 to 3,000 hours/year of solar radiation (BARBOSA, 1997).

6.4.2 Trap installation

Bucket traps containing the food attractant sugar cane were used for mass collection of M. hemipterus. The traps were distributed evenly every 45 metres. The traps were made up of 15-litre plastic buckets with perforated bottoms to facilitate drainage of rainwater. A hole was drilled in the lid of the bucket and a 10 cm diameter plastic funnel was placed in its ventral part without the narrow tube removed to allow the insects to pass into the bucket.

Fifteen sugar cane stalks (20 cm long and crushed to facilitate the volatilisation of the odour) were placed inside each trap in order to attract the insects.

Every fortnight, from September 2013 to August 2014, the individuals of M. hemipterus were removed from the traps and placed in jars with 70% alcohol. On this occasion, the sugar cane stalks were also replaced.

6.4.3 Analysing data

After tabulating the data, the spatial distribution of M. hemipterus was studied by means of geostatistics, using the semivariogram, adjusting one of the 4 possible models, Gaussian, spherical, power or exponential.

The first stage of the geostatistical study was to obtain the semivariograms, estimated on the basis of the mathematical expression:

$$\gamma * (h) = \frac{1}{2\,N\,(h)} \sum_{i=l}^{N(h)} [Z(x_i) - Z(x_i + h)]^2$$

where,

$\gamma(h)$ is the semivariance estimated by a distance h;

$N(h)$ is the number of observation pairs $[Z(x_i); Z(x_i+h)]$ separated by distance h.

When calculating the semivariogram, pairs of semivariogram values [(*y* hhé*] and distances (*h*) were obtained and arranged on a scatter plot, with the *y* values being the semivariance and the x values being the distances. A model was fitted to these points.

To better define the variogram parameters, several attempts were made to fit the spherical and exponential models, using a standard distance (range) of 240 m to model the semivariograms. All the adjustments to estimate the spatial dependence of the variables used a number of pairs greater than 30, as recommended by Guerra (1988). As for the maximum distance, the same author mentions that in practice the variographic study area rarely exceeds half of the total field sampled.

In this study, the semivariogram model that best fitted the data was the spherical model, which is described as:

$$\gamma(h) = C_0 + C_1\left[\frac{2}{3}\left(\frac{h}{a}\right) - \frac{1}{2}\left(\frac{h}{\mathrm{a}}\right)^3\right] \rightarrow 0 < h < a$$

$$\gamma(h) = C_0 + C_1 \rightarrow h \geq a$$

- Co is the nugget effect or minimum semivariance;
- Co+ CI is the plateau or maximum semivariance;
- a is the range or radius of aggregation.

All the analyses followed the procedures described by Vieira et al. (1983), who used information on the position of the sample and the value that the variable assumed at each point. In this study, the assumed variable was (incidence of *M. hemipterus*) and the point is represented by the trap. In this way, the value of the variable and the coordinates (latitude and longitude) were obtained from each sampling point and used to construct maps of *M. hemipterus* incidence using the kriging interpolation method (SILVA et al., 2011).

The parameters relating to the semivariogram were defined: the plateau (C0 + C1), the nugget effect (C0) and the range (a). Where the plateau is the value of the semivariogram corresponding to its range, i.e. from that point onwards there is no longer any spatial dependence between the samples. The Spatial Dependence Index k (ratio between C0/(C0+C1)) was also calculated to determine the spatial dependence of each evaluation according to Cambardella et al. (1994), who classify spatial dependence as weak (SDI > 75 %), moderate (25 % < SDI < 75 %) and strong (SDI < 25 %).

Both the geostatistical analysis and the population maps were drawn up using the statistical programme SURFER Version 8.0 (2002).

6.5 RESULTS AND DISCUSSION

Spatial dependence was observed among the data analysed, and the distribution of M. hemipterus occurred in aggregate form in 22 of the 24 evaluations. The spherical model was the best fit in the experimental semivariograms, and based on its conformation, the spatial dependence of the pest was verified (a), indicating the formation of plots with a radius of 78.0 to 199 m (Figures 1 and 2). In two evaluations, the spatial distribution of M. hemipterus adults occurred randomly, as can be seen in samples 11ª and 14ª, which corresponded to the following dates: 01/02/2014 and 15/03/2014.

It was observed in the 24 evaluations that the best fitted model was the spherical one, which is a strong indication that the infestation of M. hemipterus occurs through neighbouring plants and grows until it occupies the entire plantation. The aggregate distribution established by the spherical model has been the most common in insect studies (FARIAS et al., 2004), and is characterised by the formation of "clusters" in the field where these insects group together (SILVA et al., 2011).

Figures 1 and 2 show the semivariograms of the spatial distributions of M. hemipterus. The coefficient of determination (R^2) indicates the quality of the fit of the semivariogram model. Values close to one indicate a good model fit (SILVA et al., 2011), which was observed in this study, where this parameter ranged from 0.94 to 0.99.

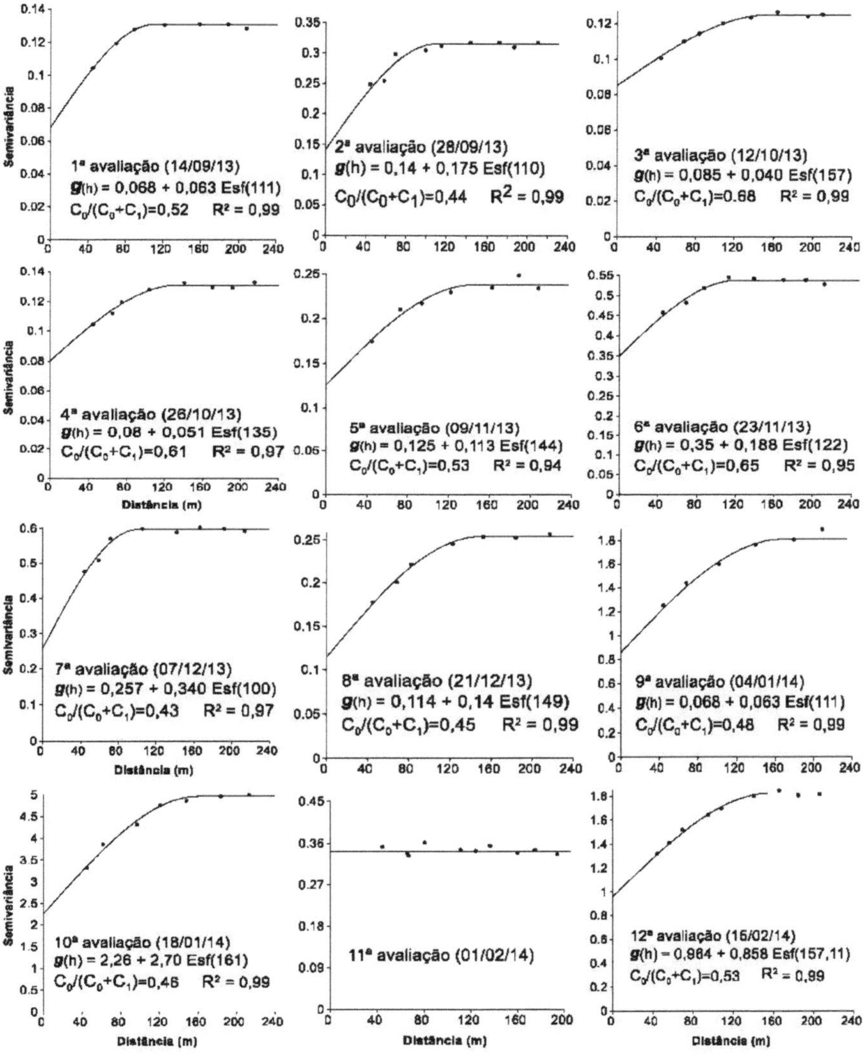

Figure 1. Semivariograms of the spatial distribution of *M. hemipterus* in the experimental area. São João da Baliza, RR, from September 2013 to February 2014.

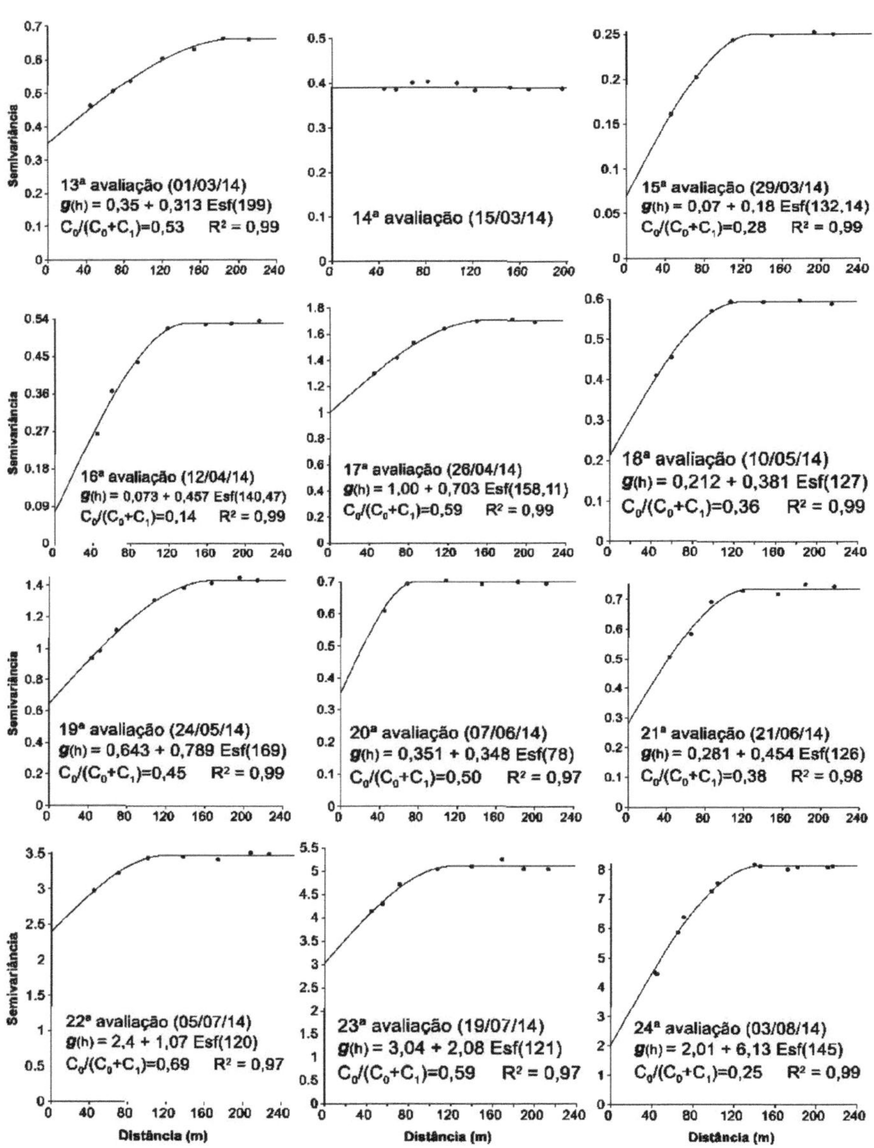

Figure 2: Semivariograms of the spatial distribution of *M. hemipterus* in the experimental area. São João da Baliza, RR, from Mar/2014 to Aug/2014.

The maximum range found in this study was 199 metres with an area of influence of 12.4 ha and the minimum was 78 metres with an area of influence of 1.9 ha. The use of the sampling grid

(45x45m), , proved to be adequate and made it possible to correctly detect the spatial distribution of M. hemipterus, since the minimum range found was 78 metres. With these results, it can be said that in order to obtain a reliable estimate of the pest population, it is necessary to have one trap every 1.9 ha, starting from the smallest range value (78 m), systematically distributed at the edges of the plantation or one trap every 78 m in the oil palm plantation. The range (a) represents the distance at which there is spatial dependence between samples, and its value has been used to determine the appropriate distance between sampling points (VALERIANO and PRADO, 2001; SHAO-KUI et al., 2005; SILVA et al., 2011).

The aggregated distribution best represented the behaviour of M. hemipterus in the experimental area. This type of distribution was also found in studies by Silva et al. (2011) analysing the spatial distribution of Aleurocanthus wogllumi in citrus in the state of Pará, Monteiro et al. (2013) determining the spatial distribution pattern of Quesada gigas nymphs in paricá plantations also in the state of Pará, Rios et al. (2014) assessing the spatial distribution pattern of Spodoptera frugiperda in maize crops in the state of Pernambuco and Pinho (2015) assessing the spatial distribution of Rhynchophorus palmarum in a palm oil plantation in the state of Pará.

According to Hall and Branham (2008) and Ray et al. (2009), in certain ecological niches it is common for populations of phytophagous insects to distribute themselves in aggregate form. This type of behaviour can bring benefits throughout an insect's life history, such as: encounters between sexual partners during reproduction, defence against natural enemies and greater exploitation of resources (HUNTER, 2000; COSTER-LONGMAN et al., 2002).

The $C0/(C0+C1)$ ratio provides a measure to estimate the degree of dependence in the samples (clustering), i.e. how much randomness there is in the surveys. The values observed in this study ranged from 0.14 to 0.69 between neighbouring points on the various sampling dates (Figures 4 and 5). These values are within the range recommended by Journel and Hijbregts (1978), who state that values below 0.80 indicate an aggregated distribution of the variable and values greater than 0.80 indicate that the phenomenon studied is tending towards randomness and that there is no spatial dependence between the samples.

Spatial dependence can be classified as strong in 2 evaluations because, in these samples, the value of the nugget effect was less than 25% of the value of the plateau and moderate in 20 evaluations because the value of the nugget effect was greater than 25% and less than 75% of the value of the plateau (CAMBARDELLA et al.,1994).

The kriging maps made it possible to see the behaviour of the spatial distribution of M. hemipterus adults and the locations of the pest's highest and lowest infestation in the oil palm plantation area (Figures 3 to 6). It was initially observed that there were a few outbreaks of M. hemipterus in the area and that, later on, plots of the pest formed on the edges of the plantation. The size of these plots was determined according to the time of the highest or lowest incidence of M. hemipterus adults (Figures 3 to 6).

Thus, in the evaluations that showed the highest population peaks of M. hemipterus adults, there were only a few patches, located on the edges of the plantation. This occurred in the following samples: 20ª, 21[(a)]and 23ª, which correspond to the evaluation periods between June/2014 and July/2014 (Figure 6). The population peaks in the aforementioned months coincided with the rainy season in the southern region of the state of Roraima. Correia (2012), working with the population fluctuation of M. hemipterus in oil palm plantations in the south of the state of Roraima, found that this insect showed higher population peaks in the months of August, September and December.

When the incidence of M. hemipterus in the area was lower, these individuals were distributed in aggregate form, but no formation of clusters was observed, as can be seen in samples 1a to 19a, 22a and 24a (Figures 3 to 6). It was found that the lowest population densities of M. hemipterus occurred from August 2013 to May 2014, months in which there was less rainfall. It was also observed that the initial infestation of M. hemipterus occurs mainly at the edges of the plantation, where the highest concentrations of these insects are found (Figures 3 to 6).

Looking at the maps, you can see the aggregate pattern of M. hemipterus infestation, which in most of the evaluations are specific locations repeated at each collection. The vast majority are on the periphery of the experimental area. One repeated aggregation point occurs on the south-west side, 50 metres from the end of the experimental area where the plantation boundary with the forest is located (Figures 3 to 6).

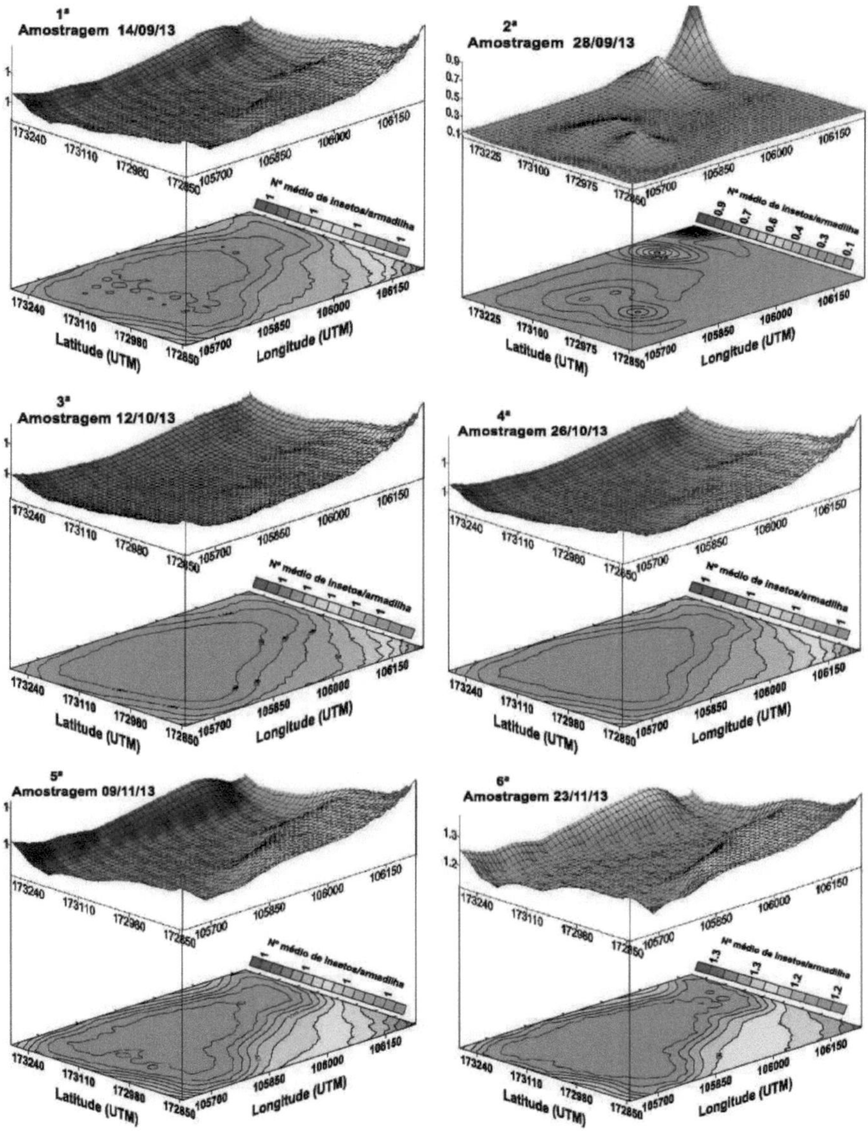

Figura 3. Maps of the spatial distribution of *M. hemipterus*, captured in the traps, from Sept/2013 to Nov/2013 in a palm oil plantation. São João da Baliza, RR.

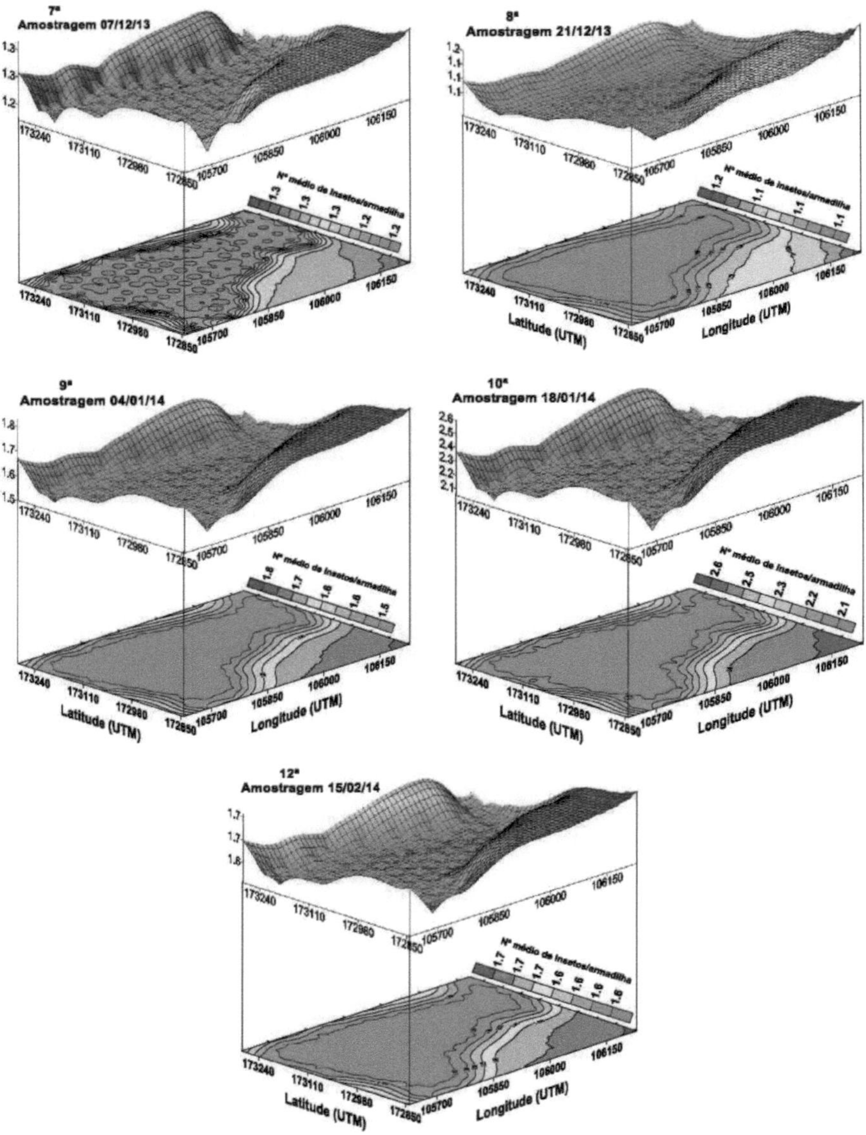

Figura 4. Kriging maps of the spatial distribution of *M. hemipterus*, captured in the traps, from Dec 2013 to Feb 2014 in a palm oil plantation. São João da Baliza, RR.

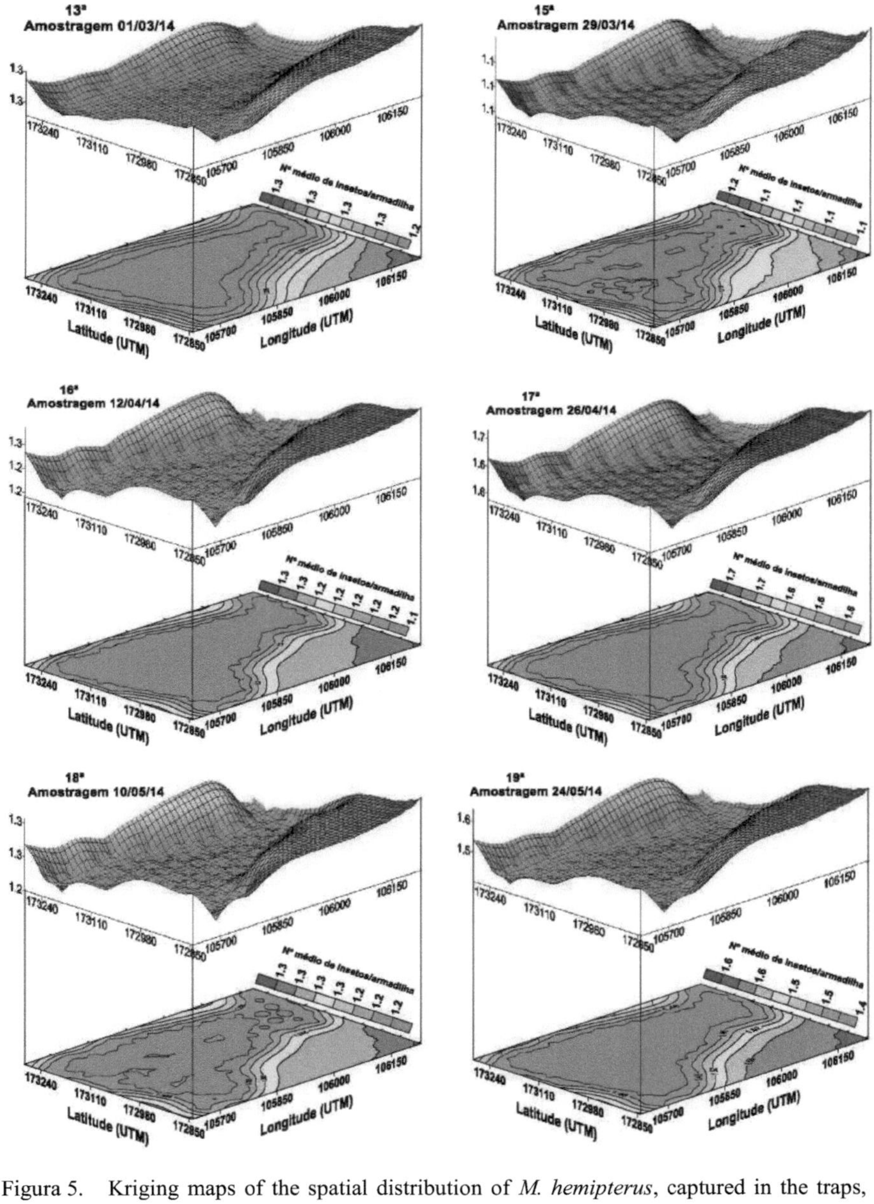

Figura 5. Kriging maps of the spatial distribution of *M. hemipterus*, captured in the traps, from March 2014 to May 2014 in a palm oil plantation. São João da Baliza, RR.

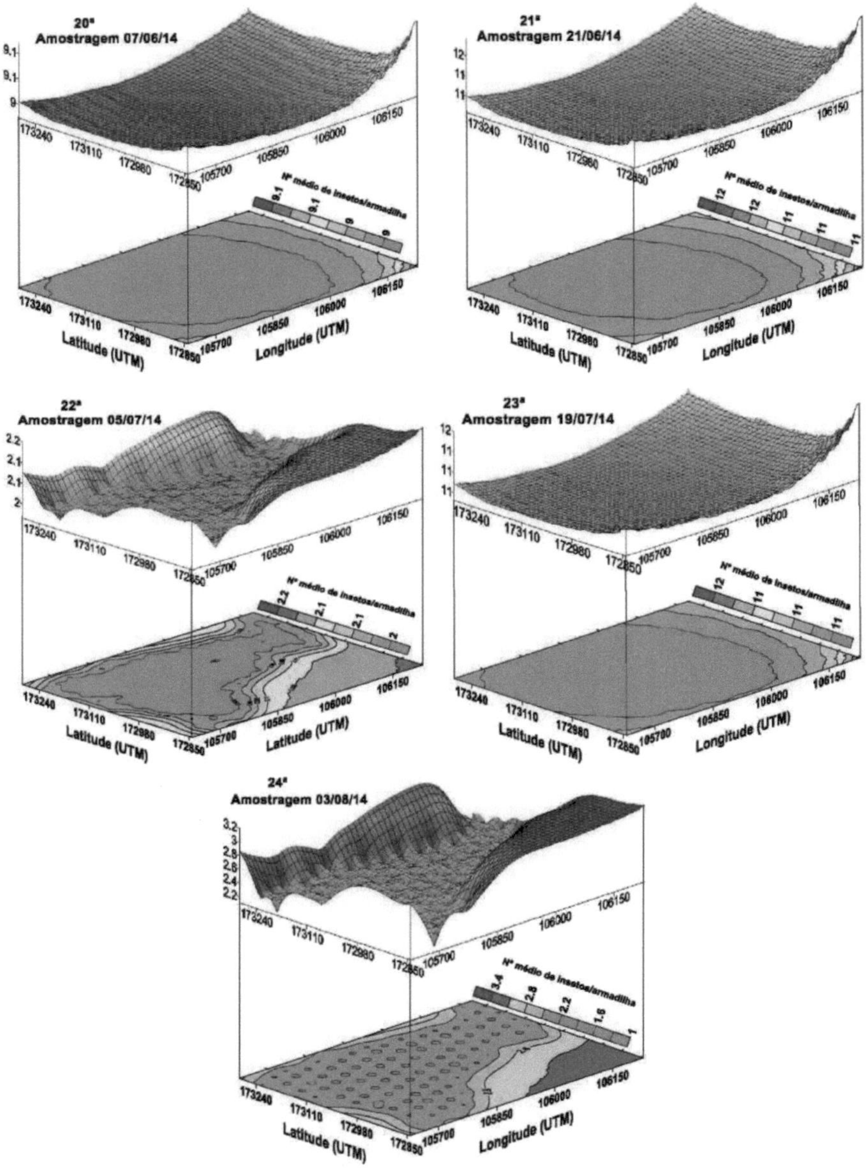

Figura 6. Kriging maps of the spatial distribution of *M. hemipterus*, captured in the traps, from June 2014 to August 2014 in a palm oil plantation. São João da Baliza, RR.

6.6 CONCLUSIONS

The spatial distribution of Metamasius hemipterus in the study area was aggregated, with spatial dependence described by the spherical model, forming "plots" between 78 and 199 metres in radius.

The initial infestation of M. hemipterus occurs predominantly at the edges of the plantation.

To develop safe sampling methods for M. hemipterus, it is recommended to distribute one trap every 12.4 ha, located on the periphery of the area.

6.7 GENERAL CONCLUSIONS

The spatial distribution of R. palmarum and M. hemipterus occurred in aggregate, with the spherical model being the best fit to the semivariograms.

For R. palmarum and M. hemipterus, the aggregation radius varied from 100 to 210 m and 78 to 199 m, respectively.

R. palmarum tends to infest different places in the oil palm crop without a specific initial location. M. hemipterus, on the other hand, congregates mainly at the edges of the plantation, without moving much towards the centre of the area.

In order to develop safe sampling methods, it is recommended to distribute one trap every 12 ha, with R. palmarum traps distributed evenly every 200 m in the plantation and M. hemipterus traps placed at the edges of the plantation.

6.8 BIBLIOGRAPHICAL REFERENCES

ABDALLA, A. L.; SILVA FILHO, J. C.; GODÓI, A. R.; CARMO C. A.; EDUARDO, J. L. P. Utilisation of by-products from the biodiesel industry in ruminant feed. Revista Brasileira de Zootecnia. v.37, p. 258-260. 2008.

AMBROGI, B. G.; VIDAL, D. M.; ZARBIN, P. H. G.; ROSADO-NETO, G. H. Aggregation pheromones in curculionidae (Insecta: Coleoptera) and their taxonomic implication. Química Nova, v. 32, n. 8, p. 2151-2158, 2009.

BARBOSA, R. I. Rainfall distribution in Roraima. In: INPA. Man, Environment and Ecology in the State of Roraima. Manaus, 1997. p. 325-335.

BRAZILIO, M.; BISTACHIO, N. J.; PERINA, V. de C. S.; NASCIMENTO, D. D. do. The oil palm *(Elaeis guineensis* Jacq.) - Review. Bioenergia em revista: diálogos, v.2, n.1, p.27-45, 2012.

CAMBARDELLA, C.A.; MOORMAN, T. B.; PARKIN, T. B.; KARLEN, D. L.; NOVAK, J. M.; TURCO, R. F.; KONOPKA, A. E. Field scale variability of soil properties in central Iowa soils. Soil Science Society of America Journal, v.58, n.6, p.1501- 1511, 1994.

CORDEIRO, A. C. C.; MACIEL, F. C. da S.; SMIDERLE, O. J.; ALVES, A. B. Desenvolvimento vegetativo de dendezeiro em ecossistemas de cerrado e floresta de Roraima. Boa Vista, Embrapa Roraima. Research and Development Bulletin, n.10, p. 16, 2009.

CORREIA, R. G. POPULATION FLUCTUATION OF Rhynchophorus palmarum, Metamasius hemipterus, AND THE NEMATOID Bursaphelenchus cocophilus IN GENDER (Elaeis guineensis) PLANTATIONS IN RORAIMA. 2012, 67 f. Dissertation (master's degree in agronomy). Federal University of Roraima.

COSTER-LONGMAN, C.; LANDI, M.; TURILLAZZI, S. The role of passive defence (Selfish Herd and Dilution Effect) in the gregarious nesting of *Liostenogaster wasps* (Vespidae, Hymenoptera, Stenogastrinae). Journal of Insect Behaviour, v.15, n.3, p.331-350. 2002.

CYSNE, A. Q.; CRUZ, B. A.; CUNHA, R. N. V. da.; ROCHA, R. N. C. da. Population fluctuation of Rhynchophorus palmarum (L.) (Coleoptera: Curculionidae) in oil palms in Amazonas. Acta Amazonica. v.43, n.2, p. 197-202, 2013.

DINARDO-MIRANDA, L. L.; VASCONCELOS, A. C. M.; VIEIRA, S. R.; FRACASSO, J. V.; GREGO, C. R. Use of geostatistics in the evaluation of the spatial distribution of Mahanarva fimbriolata in sugarcane. Bragantina, v.66, n. 3, p. 449-455, 2007.

DUARTE, A.G.; LIMA, I.S.; ARAÚJO JÚNIOR, J.V.; DUARTE, A.G.; ALBUQUERQUE, A.L.S.; CRUZ, M.M. Disposition of the nematode Bursaphelenchus coccophilus (Cobb) Baujard, in coconut trees with red ring disease. Revista Brasileira de Fruticultura, v. 30, n. 3, p. 622-627, sep. 2008.

FARIAS, P. R.S.; BARBOSA, J. C.; BUSOLI, A. C.; OVERAL, W. L.; MIRANDA, V. S.; RIBEIRO, S. M. Spatial analysis of the distribution of Spodoptera frugiperda (J.E. Smith) (Lepidoptera: Noctuidae) and losses in maize crop productivity using geostatistics. Neotropical Entomology, v.37, n.3, p.321-327, 2008.

FARIAS, P.R.S.; ROBERTO, S.R.; LOPES, J.R.S.; PERECIN, D. Geostatistical characterisation of the spatial distribution of Xyllela fastidiosa sharpshooter vectors on citrus. Neotropical Entomology, v.33, n.1, p. 13-20, 2004.

GOMES, S. M. S. Evaluation of Capture Systems for Metamasius hemipterus and Rhynchophorus palmarum (Curculionidae) in Plantations of Pupunha (Bactris gasipaes) and Dendê (Elaeis guineensis) in Southern Bahia. 2008. 54 f. Dissertation (Master's in Entomology) - Federal University of Viçosa, Viçosa-MG.

GREGO, C. R.; VIEIRA, S. R.; LOURENÇÃO, A. L. Spatial distribution of Pseudaletia sequax franclemlont in triticale under no-till management. Scientia Agricola, v.63, n.4, p.321-327, 2006.

GUERRA, P. A. G. 1988. Operational geostatistics. Ministry of Mines and Energy, 145p.

HALL, D.W.; BRANHAM, M. A. Aggregation of Calopteron discrepans (Newman) (Coleoptera: Lycidae) larvae prior to pupation. Florida Entomologist, v.91, n.1, p.124- 125, 2008.

BRAZILIAN INSTITUTE OF GEOGRAPHY AND STATISTICS - IBGE. Technical manual of Brazilian vegetation. 2. ed. Rio de Janeiro: 2012. 275p.

JOURNEL, A.G.; HIJBREGTS, C. J. 1978. Mining geostatistics. London: Academic Press, 600p.

LEÓN-BRITO, O.; VÁSQUEZ, L. N.; LÁREZ, C.; SILVA-ACUNA, R. Ciclo de vida y longevidad de Metamasius hemipterus L. (Coleoptera: Curculionidae), uma plaga de la palma aceiteira em el estado Monagas, Venezuela. Bioagro, v. 17, n. 2, p. 115-118, 2005.

MARUYAMA, W. I,; BARBOSA, J. C,; TOSCANO, L. C. Spatial distribution of Oncometopia facialis (Signoret) (Hemiptera: Cicadellidae) in a citrus orchard. Neotropical Entomology, v. 35, n.1, p. 93-100, 2006.

MOLIN, I. L D.; BARRETO, M. R. Occurrence and control of Curculionidae on Cocos nucifera L. in Sinop, Mato Grosso. Semina: Biological and Health Sciences, v. 33, n. 1, p. 53-64, 2012.

MONTEIRO, O. M.; LUNZ, A. M.; AZEVEDO, R. de; OLIVEIRA JÚNIOR, M. C. M.de. Spatial distribution and gallery depths of Quesada gigas nymphs in parica plantations. Revista Ciencias Agrarias. v. 56, n. 4, p. 353-358, 2013.

NICKELE, M. A.; OLIVEIRA, E. B de, FILHO, W. R.; IEDE, E. T,; RIBEIRO, R. D. Spatial Distribution of Anthills of Acromyrmex crassispinus (Forel) (Hymenoptera: Formicidae) in Pinus taeda Plantations. Neotropical Entomology, v. 39, n. 6, p. 862-872, 2010.

OLIVEIRA, F. Q.; MALAQUIAS, J. B.; FERREIRA, L. L.; RIBEIRO, T. da S.; ANICETO, R. R. Field evaluation of the compatibility between molasses and Beauveria bassiana in the control of Metamasius hemipterus horn, 1873 (Coleoptera: Curculionidae). Revista Verde de Agroecologia e Desenvolvimento Sustentável, v. 5, n. 4, p. 185-189, 2010.

PINHO, R. C. Spatial Distribution of Rhynchophorus palmarum (Linnaeus, 1758) (Coleoptera: Curculionidae) in an Oil Palm Agroecosystem in Eastern Amazonia. 2015. 54 f. Dissertation (Master's Degree in Agronomy) - Federal Rural University of Amazonia.

RAY, A.M.; MILLAR, J.G.; MCELFRESH, J.S.; SWIFT, I.P.; BARBOUR, J.D.; HANKS, L. M. Male-produced aggregation pheromone of the cerambycid beetle Rosalia *funebris*. Journal of Chemical Ecology, v.35, n.1, p.96-103, 2009.

RIOS, E. S.; MARTINS, I. C. F.; NORONHA, M. P. de.; SILVA, J. A. da.; SILVA FILHO, J. G. da.; BADJI, C. A. Spatial distribution of *Spodoptera frugiperda* in the wasteland of southern Pernambuco state, Brazil. Revista Ciencias Agrarias. v. 57, n. 3, p. 297304, 2014.

ROCHA, R. B. Metamasius spp. Horn (COLEOPTERA: CURCULIONIDAE) IN HELICONIAS (ZINGIBERALES: HELICONIACEAE): monitoring, associated organisms and control tactics with Beauveria bassiana and neem-based insecticides. 2012. 77f. Dissertation (Master's in Plant Production), State University of Santa Cruz.

SHAO-KUI, G. E.; RAYMOND, I. C.; ZU-FEI, M. A.; GUANG-XUE, Z.; DIAN-MO, L. I. Spatial heterogeneity and population risk analysis of cotton bollworm, *Helicoverpa armigera*, in China. Insect Science, v.12, n.4, p.255-262, 2005.

SILVA, A. G. da.; FARIAS, P. R. S.; BOIAÇU JUNHO, A. L.; CORREIA, R. G.; SILVA, J. B. da.; RODRIGUES, N. E. L. Spatial analysis of the citrus black fly in a citrus orchard using geostatistics. Revista Agricultura, v.86, n.2, p.102-114, 2011.

SOLIMAN, E. P.; PAVARINI, R.; GARCIA, V. A.; LIMA, R. C.; NOMURA, E. S.; PAVARINI, G. M. P.; DAMATTO JUNIOR, E. R. Different baits for monitoring the population of *Metamasius* sp. (Coleoptera: Curculionidae) in the cultivation of the pupa tree. Revista Eletrônica de Agronomia, v. 16, n. 2, p. 1-6, 2009.

SURFER version 8. Maps and semivariograms, New York: Golden Software, 2002.

TOLEDO, F. R.; BARBOSA, J. C.; YAMAMOTO, P. T. Spatial distribution of *Toxoptera citricida* (Kirkaldy) (Hemiptera: Aphididae) in citrus cultivation. Revista Brasileira de Fruticultura, v.28, n. 2, p.194-198, 2006.

UNTER, A.F. Gregariousness and repellent defences in the survival of phytophagous insects. Oikos, v.91, n.1, p. 213-224, 2000.

VALERIANO, M.M.; PRADO, H. Geoprocessing and sampling techniques for mapping anisotropic soil attributes. Revista Brasileira de Ciência do Solo, v.25, n.4, p.997-1005, 2001.

VIANNA, F. C. Eco-efficiency Analysis: Evaluation of the Economic-Environmental Performance of Biodiesel and PetroDiesel. 2006. 182 f. Dissertation (Master's in Chemical Engineering), Polytechnic School, University of São Paulo.

VIEIRA, S.R.; HATFIELD, J.L.; NIELSEN, D.R.; BIGGAR, J.W. Geostatistical theory and application to variability of some agronomic properties. Hilgardia, v.51, n.3, p.1- 75, 1983.

YASUDA, M. E. Survey of potential pests associated with the cultivation of the Australian royal palm Archontophoenix spp. (Arecaceae) in the Middle Itajaí Valley. 2005. 94f. Dissertation (Master's in Environmental Engineering) - Regional University of Blumenau - FURB, Blumenau-SC.

ZORZENON, F.J.; BERGMANM, E.C.; BICUDO, J.E.A. First Occurrence of Metamasius hemipterus (LINNAEUS, 1758) and Metamasius ensirotris (GERMAN, 1824) (Coleoptera, Curculionidae) on Palm Trees of the Genera Euterpe and Bactris (Arecaceae) in Brazil. Archives of the Biological Institute, v. 67, n. 2, p. 265-268, 2000.

Appendices

Table 1. Parameters of the semivariograms of Rinchophorus palmarum samples adjusted to the models

Sampling date	Semivariogram parameters*			Range (ha)[1]	Spatial dependence[2]	Model
	C_0	C_1	a (m)			
14/09/2013	0,200	0,425	206,0	13,3	Moderate	Spherical
28/09/2013	0,150	0,120	133,0	5,6	Moderate	Spherical
12/10/2013	0,035	0,088	186,0	10,9	Moderate	Spherical
26/10/2013	0,130	0,100	100,1	3,1	Moderate	Spherical
09/11/2013	0,075	0,072	159,0	7,9	Moderate	Spherical
23/11/2013	0,340	0,270	110,0	3,8	Moderate	Spherical
07/12/2013	0,301	0,524	180,5	10,2	Moderate	Spherical
21/12/2013	0,349	0,282	210,0	13,9	Moderate	Spherical
04/01/2014	0,120	0,240	165,0	8,5	Moderate	Spherical
18/01/2014	0,181	0,300	112,9	4,0	Moderate	Spherical
01/02/2014	0,360	0,225	161,6	8,2	Moderate	Spherical
15/02/2014	0,130	0,177	102,6	3,3	Moderate	Spherical
01/03/2014	0,070	0,170	155,0	7,5	Moderate	Spherical
15/03/2014	0,170	0,188	133,0	5,6	Moderate	Spherical
29/03/2014	0,080	0,290	155,9	7,6	Strong	Spherical
12/04/2014	0,380	0,264	137,7	6,0	Moderate	Spherical
26/04/2014	0,186	0,180	108,1	3,7	Moderate	Spherical
10/05/2014	0,235	0,370	145,0	6,6	Moderate	Spherical
24/05/2014	0,235	0,142	185,5	10,8	Moderate	Spherical
07/06/2014	0,320	0,190	128,9	5,2	Moderate	Spherical
21/06/2014	0,180	0,124	119,0	4,4	Moderate	Spherical
05/07/2014	0,100	0,233	100,0	3,1	Moderate	Spherical
19/07/2014	0,220	0,161	167,0	8,8	Moderate	Spherical
03/08/2014	0,199	0,208	111,6	3,9	Moderate	Spherical

*Nugget effect (c_0), spatial variance or contribution (c_1), range (a), [1]area calculated by $\pi.r^2$, where π =3.14 and r = a, (1 ha = 10,000 m²)
[2]ratio between $c_0/(c_0+c_1)$.

Table 2. Parameters of the semivariograms of Metamasius hemipterus samples fitted to the models

Date of sampling	Semivariogram parameters *			Range area (ha)[1]	Spatial dependence[2]	Model
	Co	Ci	a (m)			
14/09/2013	0.06	0.06	111.0	3,9	Moderate	Spherical
28/09/2013	0.14	0.17	110.0	3,8	Moderate	Spherical

12/10/2013	0.08	0.04	157.0	7,7	Moderate	Spherical
26/10/2013	0.08	0.05	135.0	5,7	Moderate	Spherical
09/11/2013	0.12	0.11	144.0	6,5	Moderate	Spherical
23/11/2013	0.35	0.18	122.0	4,7	Moderate	Spherical
07/12/2013	0.25	0.34	100.0	3,1	Moderate	Spherical
21/12/2013	0.11	0.14	149.0	7,0	Moderate	Spherical
04/01/2014	0.86	0.95	168.0	8,9	Moderate	Spherical
18/01/2014	2.26	2.70	161.0	8,1	Moderate	Spherical
01/02/2014	--	--	--	--	--	No adjustment
15/02/2014	0.96	0.85	157.1	7,8	Moderate	Spherical
01/03/2014	0.35	0.31	199.0	12,4	Moderate	Spherical
15/03/2014	--	--	--	--	--	No adjustment
29/03/2014	0.07	0.18	132.1	5,5	Moderate	Spherical
12/04/2014	0.07	0.45	140.4	6,2	Strong	Spherical
26/04/2014	1.0	0.70	158.1	7,8	Moderate	Spherical
10/05/2014	0.21	0.38	127.0	5,1	Moderate	Spherical
24/05/2014	0.64	0.78	169.0	8,9	Moderate	Spherical
07/06/2014	0.35	0.34	78.0	1,9	Moderate	Spherical
21/06/2014	0.28	0.45	126.0	5,0	Moderate	Spherical
05/07/2014	2.4	1.07	120.0	4,5	Moderate	Spherical
19/07/2014	3.04	2.08	121.0	4,6	Moderate	Spherical
03/08/2014	2.01	6.13	145.0	6,6	Strong	Spherical

*Nugget effect (c_0), spatial variance or contribution (c_1), range (a), [1]area calculated by $\pi.r^2$, where π= =3.14 and r = a, (1 ha = 10,000 m^2)
[2]ratio between $c_0/(c_0+c_1)$.